Multifunctional MIMO Antennas

Multifunctional MIMO Antennas

Fundamentals and Applications

Edited by

Yadwinder Kumar
Shrivishal Tripathi
Balwinder Raj

CRC Press
Taylor & Francis Group
Boca Raton London New York

CRC Press is an imprint of the
Taylor & Francis Group, an informa business

First edition published 2022
by CRC Press
6000 Broken Sound Parkway NW, Suite 300, Boca Raton, FL 33487-2742

and by CRC Press
4 Park Square, Milton Park, Abingdon, Oxon, OX14 4RN

CRC Press is an imprint of Taylor & Francis Group, LLC

ISBN: 978-1-032-19034-1 (hbk)
ISBN: 978-1-032-26857-6 (pbk)
ISBN: 978-1-003-29023-0 (ebk)

DOI: 10.1201/9781003290230

Typeset in Sabon LT Std
by KnowledgeWorks Global Ltd.

Contents

Preface

Antenna design has experienced a dramatic transformation with advancements in the technology. New analytical and modeling techniques have improved our understanding of antenna circuit design. Hence, to address the recent developments as well as discuss the futuristic aspects of antenna design, this book has come into being. It will help readers to understand both the fundamentals and the intricacies. Few concepts are discussed in more than one chapters based on area of interest as well as topic selection. The material in this book can be covered in one quarter or one semester, based on student understanding and instructor preference.

Chapter 1 discusses the recent advancements in the design of multiple-input multiple-output (MIMO) antennas owing to the rapid growth in wireless technology like 4G, 5G, and the internet of things (IoT). Moreover, fractal antennas may be useful for providing solutions to the technological advancements as they have various advantageous features such as multi-band/wideband characteristics, and compact size. In this chapter, various types of popular fractal geometries, their implementation in MIMO antennas, and their impacts on parameters like bandwidth, mutual coupling, envelope correlation coefficient, and efficiency are vividly described.

Chapter 2 deals with the antenna design that has resulted from decades of improvement and advances in wireless communication technologies. In this chapter, the iterative function system (IFS), a very useful mathematical tool, is used to describe complex fractal geometries and novel fractal shapes. Several commonly used fractals and their modified versions like Koch, Sierpinski, and Pythagorean tree fractal are discussed along with their IFSs.

Chapter 3 deals with duplex antennas have a significant role in the area of wireless communication, as they double the spectral efficiency, which is essential for 4G and 5G standards for wireless communications. Mutual coupling is the main disadvantage that disturbs the gains and radiation patterns of the radiating elements. A few techniques to reduce mutual coupling as well as the applications of the duplex antenna in various domains are discussed. in the.

Chapter 4 presents the state of the art of microstrip patch antennas and their various structures. The basic geometry of the microstrip patch antenna and its characteristics for different structures are discussed. Different feeding techniques and proximity-coupled feeding are also explained along with the mathematical analysis. The performance of wireless communication is governed by the strength and quality of the signal and by efficiency.

Chapter 5 explores how multiband MIMO antenna solutions are useful for the modern wireless communication system and why they are required. It provides a brief introduction to the multiband MIMO antenna, and characteristic mode analysis is used to provide physical insight into the structure. The demand to incorporate multiple antennas in user devices containing numerous wireless applications is increasing

Chapter 6 gives a brief idea about the design, progress, and choices of antenna elements to be used in an ultra-wideband MIMO application. The performance enhancement techniques and band notch aspects in the design are examined.

Chapter 7 presents a four-element wave patch multiband MIMO antenna design. The four elements have been placed orthogonally with respect to each other for better isolation. The antenna resonating band lies in the pioneer spectrum of the 5G band (licensed and unlicensed/shared) and wireless local area network (WLAN) n and j bands, making the antenna suitable for 5G, WLAN, and Citizens Band Radio Service usage.

Chapter 8 provides a study of the new 5G technology along with the cause-and-effect relationship between radiation and 5G. The effect of 5G on Industry 4.0 and the economy is studied. The interdependence between 5G networks and artificial intelligence is explored in the context of consumers and service providers. Moreover, the impact of COVID-19 on 5G networks and their deployment is also studied.

Chapter 9 deals with the problems/challenges associated with designing planar printed MIMO antennas using both the shared radiator and the multiple single-antenna elements approaches. Depending on the application, both linearly polarized and circularly polarized MIMO antennas can be designed. In some applications, there is a need for dual circularly polarized MIMO antennas capable of radiating both left-handed circularly polarized and right-handed circularly polarized waves.

Chapter 10 presents a four-port dual-band MIMO antenna for Wi-Fi and WLAN applications. The design comprises two planar F-shaped monopole antennas at the extremes and two elliptical-shaped patch antennas near the middle.

Chapter 11 offers brief application details of MIMO related to advancements in different technologies. MIMO-based radio-frequency identification, fabric printing for biomedical usage, WLAN, and IoT applications are discussed along with their implementation challenges. Analog, digital, and hybrid beamforming techniques are discussed with an analysis of cumulative distribution functions and energy efficiency.

Editors

Yadwinder Kumar is Senior Assistant Professor in the ECE Department, YDOE, Punjabi University Patiala, Talwandi Sabo Campus, Punjab. He earned BTech in electronics and communication engineering at GNDEC, Ludhiana in 2001; MTech in microelectronics at Panjab University, Chandigarh in 2005; and PhD at SLIET Longowal in 2016. He also served as Executive Engineer (IT Division) at Rolta India Ltd (USA MNC) Mumbai in 2005 and was selected as Officer IT (Information Technology) at PNB (Punjab National Bank). Dr. Kumar is actively working in the field of design and development of planar hybrid fractal antennas and antennas for IoT and RFID.

He also serves as a reviewer with highly renowned international journals such as *Wireless Personal Communication* (Springer), *International Journal of Electronics* (Taylor & Francis), *Applied Computational Electromagnetic Society Journal* (ACES), *Journal of Nanoelectronics and Optoelectronics* (American Scientific Publisher), *International Journal of Microwave and Wireless Technologies* (Cambridge), *Journal of Medical Systems* (Springer), *International Journal of Electronics Letters* (Taylor & Francis), *Indian Journal of Pure and Applied Physics* (NIScPR*),* *International Journal of Computing and Digital Systems (IJCDS)*, *International Journal of Theoretical Physics (Springer)*, and *Health and Technology Journal* (Springer).

Dr. Kumar has published papers in international journals of high repute, attended various IEEE and other international conferences, and also authored book chapters. Presently, he is working on inkjet printed and 3-D printed antennas for various wireless applications. Dr. Kumar is a lifetime member of the Indian Microelectronics Society (IMS). He has delivered a large number of expert lectures in various short-term courses, Faculty Development Programme at the department and university level.

Shrivishal Tripathi is an Assistant Professor at International Institute of Information Technology, Naya Raipur. Prior to this, he was associated with BITS-Pilani, Hyderabad and NIIT University, Neemrana as an Assistant Professor. He earned a PhD at the Indian Institute of Technology Jodhpur in the Electrical Engineering Department. He earned an ME (electronics and electrical communication) at PEC University of Technology, Chandigarh in 2011. He earned a BTech (electronics and communication engineering) at Dr. K. N. Modi Institute of Engineering and Technology, Ghaziabad in 2009.

His main research interests include UWB antennas, MIMO antennas, reconfigurable antennas, antennas for IoT, antennas for 5G, SAR analysis, FMCW radar, and RF/microwave circuit design. He is serving as the Principal Investigator (PI) for two project sanctioned by the Department of Science and Technology (DST), Govt. of India and Chhattisgarh Council Of Science And Technology (CCOST) that is in progress. He has also served as co-PI for two completed projects sanctioned by Indira Gandhi Krishi Vidyalaya (IGKV), Raipur. He has published many papers in reputed journals such as *IEEE Antenna and Wireless Propagation Letters*, *IEEE Canadian Journal of Electrical and Computer Engineering, Springer Wireless Personal Communication*, *IET Antenna and Propagation*, and Wiley's *Microwave and Optical Technology Letters*. Moreover, he has presented several papers at IEEE international/national conferences held in India and abroad. He has organized many workshops/Faculty Development Programmes in the areas of RF/microwave and communication. He is also the coordinator of the IIRS-ISRO outreach program. In addition to that contributed in "National Rogue Drone Policy" of Government of India.

Balwinder Raj is an Associate Professor in the ECE Department at NITTTR Chandigarh. He previously worked at the National Institute of Technology Jalandhar, Punjab, India (2012–2019). He earned a BTech in electronics engineering at PTU Jalandhar, an MTech in microelectronics at PU Chandigarh, and a PhD in VLSI Design at IIT Roorkee, India, in 2004, 2006, and 2010, respectively. For further research work, the European Commission awarded him a mobility of life research fellowship for postdoc research work at the University of Rome, Tor Vergata, Italy (2010–2011). He was a visiting researcher at KTH University Sweden (October–November 2013). He has visited Japan, Australia, Malaysia, Italy, Sweden, Finland, and Thailand for research collaborative projects, invited lectures, and conferences.

Dr. Raj received the best teacher award at ISTE New Delhi in July 2013 and the Young Scientist Award from Punjab Academy of Sciences in February 2015. He also received a best research paper award at the international conference on electrical and electronics engineering held at Pattaya, Thailand, in July 2015 and an early carrier research award from SERB-DST Government of India in 2017.

He has authored or coauthored 4 books, 8 book chapters, and more than 90 research papers in peer-reviewed international/national journals and conferences. He has guided seven PhDs, and currently four PhD scholars are working with him. He has guided more than 30 MTech students. Dr. Raj has completed three research projects from DST Delhi, SERB Delhi, and CIMO Finland. Currently, he is handling two research projects from SERB Delhi and ISRO. His areas of interest in research are classical/non-classical nanoscale semiconductor device modeling, FinFET-based memory design, low-power VLSI design, digital/analog VLSI design, and FPGA implementation.

Contributors

Santanu Kumar Behera
Professor
Department of Electronics and Communication Engineering
National Institute of Technology Rourkela
Rourkela, India

Atanu Chowdhury
Assistant Professor
Department of Electronics and Communications Engineering
Seacom Engineering College
Maulana Abul Kalam Azad University of Technology
West Bengal, India

Tanmaya Kumar Das
Assistant Professor
Department of Electronics and Communication Engineering
C.V. Raman Global University
Bhubaneswar, India

Santanu Dwari
Associate Professor
Department of Electronics Engineering
Indian Institute of Technology (IIT-ISM)
Dhanbad, India

Biswajit Dwivedy
Assistant Professor (Senior Grade-1)
School of Electronics Engineering (SENSE)
Vellore Institute of Technology
Vellore, India

Prabir Ghosh
Assistant Professor
Department of Electronics and Communications Engineering
Seacom Engineering College
Maulana Abul Kalam Azad University of Technology
West Bengal, India

Binod K. Kanaujia
Professor
School of Computational and Integrative Sciences
Jawaharlal Nehru University (JNU)
Delhi, India

Harleen Kaur
PhD Research Scholar
Department of Electronics and Communication Engineering
Thapar Institute of Engineering and Technology
Patiala, India

Manjit Kaur
Principal Engineer
Education and Training Division
C-DAC Mohali
Mohali, India

Manpreet Kaur
PhD Research Scholar
Department of Electronics and Communication Engineering
Thapar Institute of Engineering and Technology
Patiala, India

Sachin Kumar
Research Assistant Professor
SRM Institute of Science and Technology
Chennai, India

Sunil Kumar
Ex Student
Department of Electronics and Communication Engineering
Punjabi University Patiala Regional Campus
Yadavindra Department of Engineering (YDoE)
Talwandi Sabo, India

Yadwinder Kumar
Assistant Professor
Department of Electronics and Communication Engineering
Punjabi University Patiala Regional Campus
Yadavindra Department of Engineering (YDoE)
Talwandi Sabo, Punjab

Richa Kumari
Ex Student
C-DAC Mohali
Mohali, India

Tejaswita Kumari
Assistant Professor
Department of Electronics and Communications Engineering
Seacom Engineering College
Maulana Abul Kalam Azad University of Technology
West Bengal, India

Durga Prasad Mishra
PhD Research Scholar
Department of Electronics and Communication Engineering
National Institute of Technology Rourkela
Rourkela, India

Ajay Mudgil
Joint Director
Education and Training Division
C-DAC Mohali
Mohali, India

Punya P. Paltani
Assistant Professor
Dr. SPM International Institute of Information Technology
Naya Raipur, India

D. Venkata Siva Prasad
PhD Research Scholar
Dr. SPM International Institute of Information Technology
Naya Raipur, India

Balwant Raj
Assistant Professor
University Institute of Engineering and Technology (UIET)
Panjab University
SSG Regional Centre
Hoshiarpur, India

Shobhit Saxena
Assistant Professor
Department of Electronics and Communication Engineering
Chaudhary Charan Singh University Campus
Meerut (U.P), India

Shalini Shah
Assistant Professor
Department of Electronics and Communication Engineering
Amity School of Engineering and Technology
Amity University
Noida, India

Balwinder Singh
Joint Director
Academics and Consultancy Division
C-DAC Mohali
Mohali, India

Hari Shankar Singh
Assistant Professor
Department of Electronics and Communication Engineering
Thapar Institute of Engineering and Technology
Patiala, India

Harsh Verdhan Singh
PhD Research Scholar
Dr. SPM International Institute of Information Technology
Naya Raipur, India

Mandeep Singh [Chapter 2]
Joint Director
Robotics and Smart systems Division
C-DAC Mohali
Mohali, India

Mandeep Singh [Chapter 9]
PhD Research Scholar
National Institute of Technology
Jalandhar, India

Shrivishal Tripathi
Assistant Professor
Dr. SPM International Institute of Information Technology
Naya Raipur, India

Devica Verma
Student
Department of Electronics and Communication Engineering
Amity School of Engineering and Technology
Amity University
Noida, India

Chapter 1

Introduction to fractal antennas and their role in MIMO applications

Biswajit Dwivedy and Tanmaya Kumar Das

CONTENTS

1.1 INTRODUCTION

Fractal geometry is based on the process of iteratively generating contours with interminable complicated fine structures. The word *fractal* was taken from the Latin term *fractus*, which is associated with the verb *fangere*, meaning "to split or break." In 1970, B. B. Mandelbrot, a French mathematician, introduced the term *fractal* after his intense research on various irregular and fragmented geometrical shapes inspired by nature such as

DOI: 10.1201/9781003290230-1

trees, snowflakes, ferns, and leaves. Later, the fractal geometries were used for a wide range of applications in different sectors of science and engineering. At the same time, fractals were implemented in antenna engineering as a means of imposing inherent self-similarity and self-affinity in order to model the complex geometrical structures of the antennas and achieve a massive improvement in their characteristics such as size miniaturization, multiband response, broadband, impedance matching, and stable radiation properties. Therefore, fractal-shaped antennas are classified as a special type of electromagnetic radiating structure where the overall dimension is composed of a set of replications of a specific geometry and each geometrical iteration occurs on a different scale [1, 2]. Although many mathematical formulations have been developed to represent various fractal shapes, there are no strict guidelines that can be considered suitable for all fractal geometries. Fractals can be characterized based on geometrical properties. *Self-similarity* is one of these characteristics; its presence indicates the whole geometry is formed by replicating the small structure on a diminished scale. Another important characteristic is *self-affinity*; its presence indicates the total geometry is generated by scaling the pieces differently in the X and Y directions. Here, the small portions of the geometry are not identical to the complete structure; rather, they are slanted or twisted in different scales, resulting in anisotropic transformations. Recently, the design and performance of planar antennas based on various fractal geometries like Koch, Sierpinski gasket, and Minkowski were comprehensively investigated in the literature [3–7]. In many communications [8–13], specialized fractal geometry–based microstrip antennas were also proposed to meet various contemporary wireless standards.

This chapter aims to provide a sound understanding of various fractal shapes, including the techniques of introducing fractal geometry of different orders over Euclidean geometry for fascinatingly perturbing the current distribution to make the antenna invariably different from the conventional ones. It also describes the use of fractals in multiple-input multiple-output (MIMO) terminal-based antennas and their effect on various parameters like impedance bandwidth (IBW), the envelope correlation coefficient (ECC), capacity loss, mutual coupling, and the spacing between the radiators implemented for various wireless applications.

1.2 FRACTAL FOR ANTENNAS: THE TECHNIQUES AND PURPOSE

Fractal engineering comprises a class of topology-based miniaturization techniques by which characteristics like the geometry/configuration, surface current density distribution (electric or magnetic), and electrical dimensions of an antenna can be altered to improve its performance. In other words, fractals are space-filling geometries with a Hausdorff dimension that can accommodate relatively greater lengths within a much smaller area. This elemental concept, along with the self-similarity feature of fractal geometry, is effectively used by antenna engineers to create *fractal antennas*, which are superior to conventional antennas based on Euclidean geometry.

The fractal antenna can perform like a large antenna by reducing the area to a great extent. This can be understood from a simple example of a Koch dipole antenna. A Koch dipole antenna is formed by amending the straight-wire dipole antenna in an iterative process using a special mathematical formulation [14–16]. The detailed mathematical formulation of the Koch, along with those of other popular fractal shapes, is given in the next section of this chapter. As shown in Figure 1.1, the initial step of the Koch curve starts with a straight line (zeroth iteration), and the total length of the curve corresponding to the *n*th iteration is $(4/3)^n$ of the initial length. This implies that even if the Koch curve–based dipole has the same starting and ending points as the normal dipole, it has a physical length much greater than that of the ordinary one with a nether resonant frequency. The decrement of the resonant frequency of the Koch dipole with the increase in the iteration was investigated in [16]. It was highlighted that even after each iteration, where the overall length of the Koch fractal increases by 1/3 of the previous one, the resonant frequency does not decrease proportionately. This phenomenon can be explained by considering that the breaking of each fractal into smaller segments during each iteration lowers the resonant frequency while making the wavelength longer. As the number of iterations increases, at some point all the added segments become much smaller than the wavelength (segment length $\ll 0.1\ \lambda$) and also appear to be

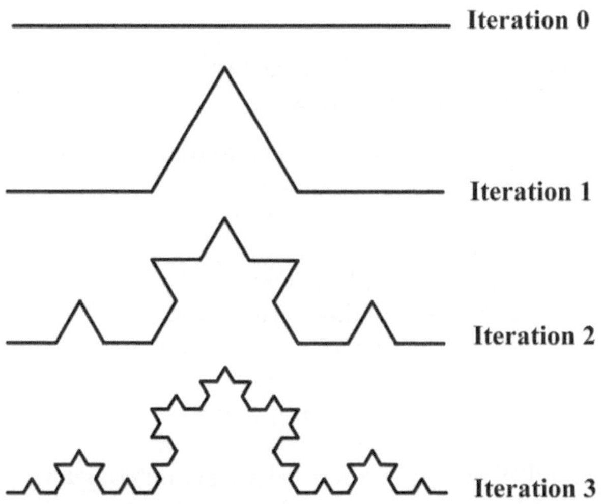

Figure 1.1 Generation of the Koch curves.

smooth curve to the current. Hence, all these small segments are undistinguished by the current and no longer reduce the resonant frequency. Even though the fractal length is incongruent with the resonant frequency of the antenna, the fractal gives effective miniaturization at the lower frequency. Along with size reduction, fractals can be used to improve antenna performance in terms of input impedance, Q-factor, gain, radiation efficiency, bandwidth, and change of polarization/pattern.

In the current scenario, all mobile devices are meant to cover a variety of wireless standards—including GSM (Global System for Mobile Communication) 800–900, GSM 1800–1900, Bluetooth (2.4 GHz), WLAN (wireless local area network, operating at 2.4, 5, and 5.8 GHz), fourth generation (4G), and fifth generation (5G)—and to operate within the defined bandwidth [17]. The most popular choice for these mobile devices is the planar inverted-F antenna (PIFA), which is not only compact or low profile but also responsive to both linearly polarized waves (whether horizontal and vertical). It has little back radiation toward the user, causing less electromagnetic absorption, according to [18]. Since fractal antennas have the ability to revamp the current distribution according to their self-similar geometry, they can have multiple resonance bands with a small form factor. This feature makes fractal antennas suitable for cellular devices. Therefore, in [18], the fractal concept was applied to the PIFA to achieve multiband support with reduced size. Also, fractal antennas can be implemented in vehicles (including autonomous cars) to provide various radio services like GSM, Bluetooth, and the Global Positioning System and can be used in various types of sensing applications. Despite many beneficial features of the fractal-based antenna, sometimes the size miniaturization is achieved at the expense of radiation efficiency.

1.2.1 Advantages and disadvantages of fractal-based antennas

Based on various applications, fractal antennas have the following advantages:

- Device size miniaturization
- Generation of multiple bands or wideband
- Improved impedance-matching performance
- Congruous radiation performance throughout the operational band

However, they suffer from the following problems:

- Fabrication difficulty of higher-order fractal geometry
- Restrictions on the repetition of the fractal geometry
- Occasional deterioration of the radiation efficiency

1.2.2 Popular fractal geometries used for antennas

There are many ways of representing the dimensions of fractal geometries, but the easiest and most popular technique is to use the scaling factor based on self-similarity. Suppose a geometry is divided into K identical copies and the originated subgeometry is diminished to a factor x of the original. Then the dimension of the fractal, D_f, can be represented as in Equation (1.1) [2, 19].

$$D_f = \frac{\log k}{\log 1/x} \tag{1.1}$$

In this section, some popular fractal geometries are succinctly described to provide readers with a background in the fundamentals.

1.2.2.1 Koch curve

The Koch curve, one of the popular fractal-based geometries, is named after Swedish mathematician Helge Von Koch [20]. The generation of the Koch curve with different iteration orders (IOs) is shown in Figure 1.1. Here, the initial length a for IO = 0 is split into three identical segments of length $a/3$ in IO = 1, and the central segment is replaced with two segments of length $a/3$, each forming one upper side of an equilateral triangle. The above process is repeated an infinite number of times for higher IO values.

Some special features characterize this fractal form. The Koch curve is a self-similar object where exactly similar versions of the curve, scaled down by a factor of three, are found within the structure. Also, the curve is nowhere differentiable; i.e., no point of the ideal curve has a defined tangent [20]. Similarly, the other striking properties belong to the length and area of the curve. Although its shape looks the same when the number of iterations is large enough, the curve increases its length at each iteration up to an infinite length for the ideal fractal shape. The length l_n^{Koch} of the nth iteration of the Koch curve is given by Equation (1.2) [20].

$$l_n^{Koch} = a(4/3)^n \tag{1.2}$$

Here, a is the length of the initial segment, and when $n \to \infty$, $l_n^{Koch} \to \infty$.

At iteration 1, the smallest element length can be expressed as $\epsilon_1 = a/3$. Now the overall length after the nth iteration can be rewritten as Equation (1.3) [21].

$$l_n^{Koch} = a^D \epsilon_n^{(1-D)} \tag{1.3}$$

where $D = \dfrac{\log 4}{\log 3} = 1.261$ and $\epsilon_n = a/3^n$.

In Equation (1.3), the total length l_n^{Koch} increases exponentially by a factor D for the initial length a with a constant value of unit length (ϵ_n). The

Figure 1.2 Iteration order and construction of Koch snowflake fractal.

exponential factor of ϵ_n—i.e., $D - 1$—is based on Richardson's law, showing the divergence of l_n^{Koch} as $\epsilon_n \rightarrow 0$ [20, 22].

1.2.2.2 *Koch snowflake*

The Koch snowflake or Koch island is a variant of fractal geometry constructed by applying the Koch curve fractal to each line segment of an equilateral triangle [20]. The generation of a Koch snowflake with different IOs is shown in Figure 1.2. The scaling law can be applied to this fractal geometry having one independent unit length ϵ_n, and Equation (1.3) can be expressed as [20, 21]

$$l_n^{Koch} / \epsilon_n = f\left(a / \epsilon_n\right) = \left(a / \epsilon_n\right)^D \quad (1.4)$$

1.2.2.3 *Cantor set*

Mathematically, a Cantor set is a set of collinear points possessing several salient features [20, 22]. It was introduced in 1883 by German mathematician George Cantor. Figure 1.3 shows the generation of the Cantor set fractal through IO = 5 by iterative deletion of the central of three individual line segments. Using Equation (1.1), the factor D for this fractal design can be written as

$$D = \frac{\log 2}{\log 3} = 0.6309 \quad (1.5)$$

The Cantor set where $0 \leq D \leq 1$ is called *dust* due to the presence of points.

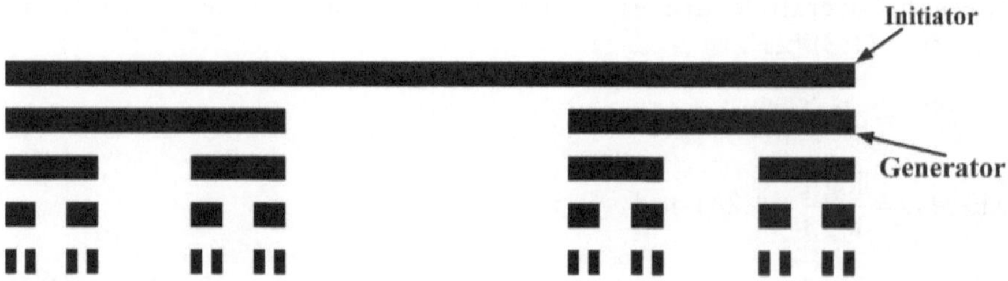

Figure 1.3 Iteration order and construction of the Cantor set fractal.

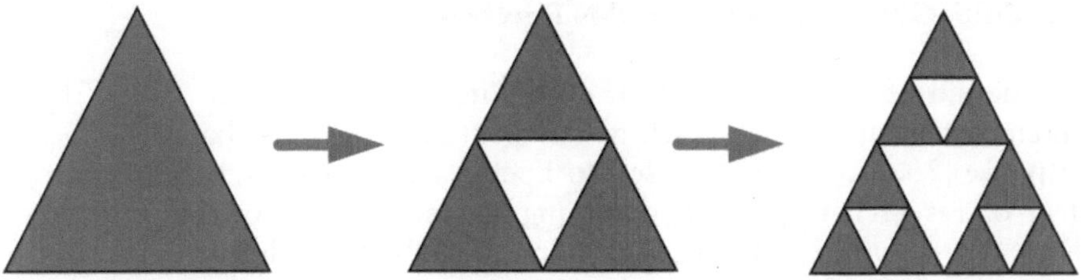

Figure 1.4 Iteration order and construction of the Sierpinski gasket fractal.

1.2.2.4 Sierpinski gasket

The Sierpinski gasket/triangle/sieve fractal was introduced by Polish mathematician Waclaw Sierpinski in 1915, and in the last two decades, it has been extensively used for antenna design for several applications [20, 22]. Sierpinski geometry is a fixed-set fractal, and its generation with different iterations is shown in Figure 1.4. The factor D can be evaluated from Equation (1.1) as

$$D = \frac{\log 3}{\log 2} = 1.585 \tag{1.6}$$

1.2.2.5 Sierpinski carpet

A revised form of the Sierpinski fractal known as the Sierpinski carpet was introduced by the same mathematician in 1916. It can also be considered as a generalization of the Cantor set to a two-dimensional form [20, 22, 23]. The generation of the Sierpinski carpet is shown in Figure 1.5. The expression for the fractal dimension D is given in Equation (1.7).

$$D = \frac{\log 8}{\log 3} = 1.8928 \tag{1.7}$$

Figure 1.5 Iteration order and construction of the Sierpinski carpet fractal.

1.3 DESIGN OF FRACTAL ANTENNAS

The design of fractal antennas is a growing field of research due to these antennas' useful features such as compact size and multiband/wideband response [20, 22]. In comparison to Euclidean geometries, there is a great deal of research interest in designing antennas using fractal geometry. Table 1.1 compares the general characteristics of Euclidean and fractal geometries [20–22].

Fractal geometry is based on an iterative process in which the parts of the structure are repeated with different scaling factors to achieve design goals [20]. This feature of fractal structures is known as the self-similarity property. Hence, fractal-based antennas can be considered to operate over multiple bands with similar radiation characteristics. Similarly, the space-filling features of fractal geometry can be regarded as beneficial for antenna design [20]. In [19, 23], the Hilbert curve was used for the construction of a space-filling fractal curve with different IO values. The space-filling feature was mainly employed to obtain compact classical antennas such as loops and dipoles. The meandering process employed in these designs can effectively fill the space, which increases the overall electrical length in a compact size [19]. Koch curves are a classic example of fractal geometries utilizing the self-similar and space-filling properties to design antenna elements. The designed antennas provide compact and multiband/wideband characteristics [21, 24]. Similarly, Sierpinski carpet fractal geometry can be employed to obtain ultra-wideband (UWB) antennas [20, 25].

Fractal structures have also been used in the design of antenna arrays, providing essential characteristics such as multiband response, reduction in side-lobe level, and development of rapid beam-forming algorithms. With these features, antenna arrays can be effective in the design of a reconfigurable system [21]. Further, this fractal-based array design technique can be customized to create antennas for MIMO terminals with the above-mentioned characteristics.

Table 1.1 Comparison of the Euclidean and fractal geometries

Feature	*Euclidean geometry*	*Fractal geometry*
Definition	(i) Formula based (ii) Analytical equations	(i) Iterative rule based (ii) Recursive algorithms
Application	Artificial objects	Natural objects
Shape variation	Changes with scaling	No changes with scaling (due to self-similarity)
Elements	Vertices, edges, surfaces	Iterations of functions
Differentiable	Yes, locally smooth	No, locally rough

1.3.1 Recent advances in the design of fractal antennas

The distinguishing features of fractal antennas such as space filling, periodicity, and self-similarity help to make the design compact, allowing better performance in terms of multiband and multimodal responses [19–21]. In many cases, the upgradation in antenna performance does not properly correlate with the attributes of fractal geometry. Hence, the development of a novel fractal antenna requires the proper conceptualization and mathematical formulation of a suitable fractal shape utilizing the features of the fractal. There is a vast range of applications where different fractal antennas can provide better performance. This subsection aims to provide insight into the versatility and broad application domain of fractal antennas, as well as their characteristics, through multiple design examples.

1.3.1.1 Wideband antennas

In many communications, it has been proven that the self-similarity and space-filling features of fractal geometry can greatly improve the bandwidth of an antenna while still occupying a small physical space [19–21]. In [26], a planar monopole antenna was introduced using Koch fractal geometry in the radiator as well as the ground plane. The measurement results show a fractional bandwidth of 122% (3.4–13 GHz) with a stable radiation pattern throughout the entire band.

Similarly, in [27], a hexagonal microstrip patch antenna (MPA) was loaded with square-shaped Sierpinski fractal slots of the second-order iteration (IO = 2) to obtain impedance matching ($|S_{11}| \leq -10$ dB) in a super-wideband (SWB) frequency range of 3.4 –37.4 GHz.

1.3.1.2 Ultra-wideband antennas

Recently, there has been a huge demand for UWB systems covering a broad range of applications, including ground-penetrating radar (GPR), WLAN, and military applications, within a wide frequency range of 3.1–10.6 GHz [25, 28–32]. For UWB systems, antennas with stable radiation patterns, constant gains, and simple structures are essential, along with wide IBW. Incorporation of the fractal can help in designing compact antennas with all these characteristics.

The UWB monopole antenna proposed in [28] consists of triangular fractals embedded within hexagonal rings and a defected ground structure (DGS). The reflection coefficient response ($|S_{11}|$) confirms that the antenna covers a bandwidth of 8.4:1 in the range from 3 to 25.2 GHz, while its gain values vary from 3 to 9.8 dBi over the entire wideband.

In [25], a compact slot antenna using Sierpinski square fractal geometry and having UWB characteristics was constructed. The antenna shows a band-notch feature as well. The UWB frequency range of 3.1–10.6 GHz was achieved by increasing the fractal iteration up to the fourth order and inserting a rectangular groove on the ground plane. Similarly, the ∩-slot on the feed line provides the required band-notch response at 5.5 GHz (5–6 GHz).

MPAs are currently very popular due to their intrinsic features, including lower cost of fabrication, smaller size, and ease of integration with electronic devices. Despite these features, MPAs have narrow bandwidth. Hence, bandwidth must be enhanced by switching to different antenna structures such as monopole [29] and Vivaldi [30]. The log-periodic technique can also be used to obtain a larger bandwidth and a stable radiation pattern [31] with a larger dimension. In all these modified MPAs, fractal designs can be incorporated to reduce their size while not hampering their wideband performance. In [32], a square fractal-based log-periodic MPA was introduced to achieve a UWB response. The measurement results indicate a high IBW of 3.1–10.6 GHz and a maximum realized gain of 9 dB at 10.6 GHz with stable radiation characteristics throughout the UWB range.

1.3.1.3 Multiband antennas

The multiband characteristic is another inherent feature of fractal antennas, along with wide bandwidth and compactness [33, 34]. A simple H-shaped fractal antenna was introduced in [35] to obtain a multiband MPA for WLAN applications. The measured responses of the antenna show multiple resonances at 2.45 GHz (2.4–2.5 GHz) and 5.5 GHz (5.0–6.0 GHz) with a peak directivity of 7.52 dBi at the higher band (5.5 GHz).

A modified Sierpinski gasket fractal–based multifrequency reconfigurable antenna was described in [36]. Frequency reconfigurability was obtained by using a variable feeding arrangement in the fractal design. Measurement validates that for different feed positions, the antenna can be operated in two single-band modes (at 1.1 GHz and 3.4 GHz), in three dual-band modes (at 1.1 and 3.4 GHz, 1.1 and 5.8 GHz, and 3.4 and 5.8 GHz), or in one triple-band mode (at 1.1, 3.4, and 5.8 GHz).

1.3.1.4 Wearable antennas

There is a huge demand for wearable-antenna technology in different fields, including sports, the military, health care, and radio frequency identification (RFID). This technology has required extensive research on the substrate materials, including fabrics and polymers [37, 38]. In [39], a fractal-based antenna using jean-fabric substrate material was designed for dual-band operation. The design embeds a slotted electromagnetic bandgap (EBG) structure and can be useful for wearable applications covering GSM 1800 and 2.45 GHz industrial, scientific, and medical (ISM) bands. Additional bending and on-body measurements with specific absorption rate (SAR) analysis validate the usefulness of the antenna for wearable applications.

Similarly, a triangular-shaped MPA incorporating Koch fractal geometry, meandered slits, and DGS for a wearable on-body wireless body area network (WBAN) application was presented in [40]. The structurally conformable fractal antenna was realized using RT/Duroid 5880 substrate ($\varepsilon_r = 2.2$) and has a compact dimension of $39 \times 39 \times 0.508$ mm^3, which corresponds to $0.318\ \lambda_0 \times 0.318\ \lambda_0 \times 0.004\ \lambda_0$ at 2.45 GHz. It shows an IBW of 7.75% from 2.36 to 2.55 GHz in the ISM band with a peak realized gain of 2.06 dBi at the operating frequency of 2.45 GHz. It also shows a maximum allowable input power for the antenna of 0.34 W and 0.26 W for the arm and leg, respectively, so it has a SAR value within the safety limits defined by the European Union.

In [41], a tree-shaped fractal antenna was introduced with a smaller metallic footprint designed on a flexible substrate. The fractal antenna consists of a self repetitive V-shaped tree structure with copolanar waveguide (CPW) feeding. Measurement indicates a shift in resonance frequency from 16.38 to 5.08 GHz with a 16% increment in metallic footprint by varying the fractal iteration from $n = 1$ to $n = 4$. The antenna has a peak realized gain of 1.6 dBi at 16.38 GHz, and the gain value drops by only 2.4 dB for the higher-order iteration ($n = 4$) without affecting the size. Overall, the flexible antenna is compact, lightweight, and suitable for wearable applications.

1.3.1.5 Antennas for RFID applications

RFID technology has a growing market in various sectors such as hospitals, warehouses, supermarkets, inventory control, health care, and access control [42]. The RFID system consists of a reader/interrogator, tag, and middleware software. The tags are attached to the objects to be identified and may orient in a random direction. Therefore, circularly polarized (CP) reader antennas are essential for efficient tag reading [42].

In [43], a CP MPA with Koch fractal geometry was introduced for an ultrahigh frequency (UHF) RFID system. The design consists of two asymmetrical Koch fractals with four arrow-shaped slots. The fractal antenna resonates at 911 MHz with an IBW of 37.0 MHz (891.0–928.0 MHz) and an axial ratio bandwidth (ARBW) of 8 MHz (907.0–915.0 MHz).

Similarly, a compact square-shaped CP antenna using Koch fractal geometry at the radiating edge was presented in [44] as an RFID reader at an ISM band. Figure 1.6 shows the geometrical layout of the CP antenna. Two antenna prototypes were fabricated operating at two frequency bands—i.e., at 2.435 GHz (antenna 1) and 5.78 GHz (antenna 2). Measurement confirms a 3 dB ARBW covering 2.433–2.439 GHz and 5.75–5.8 GHz for antenna 1 and antenna 2, respectively. Also, the lower-band antenna has a maximum realized gain of 6.93 dBic at 2.435 GHz, whereas the upper-band antenna shows a gain of 6.15 dBic at 5.78 GHz.

The design in [45] is a CPW-fed fractal monopole antenna developed for short-range RFID reading applications with a maximum reading distance of 1.32 m. The antenna has an IBW of 35.7% at the resonant frequency of 910 MHz and a 3 dB ARBW of 4% varying from 900 to 936 MHz.

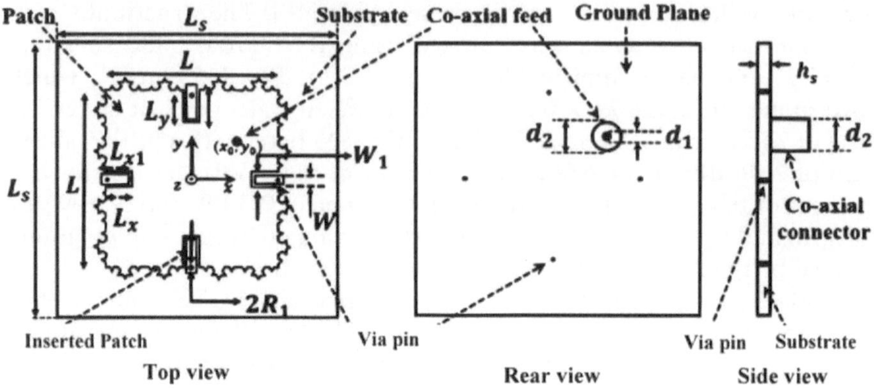

Figure 1.6 Geometry of the Koch fractal-based CP antenna for RFID reader applications. (Reprinted from [44] with permission from Elsevier.)

1.4 MIMO ANTENNAS USING FRACTAL GEOMETRY

The rapid growth in wireless sensors and 4G/5G wireless technology has resulted in a lot of research interest in the design of MIMO antennas. The wireless channel between the transmitter and receiver is inclined to multipath fading, interference, and losses due to radiation. The severity of these losses multiplies at higher frequencies. Therefore, the use of MIMO antennas becomes very significant for enhancing the range of transmission with no change in signal power. In MIMO systems, signal transmission and reception can be managed in an agile manner with the use of multiple antennas, which help to achieve low latency and maximum throughput as well as to enhance channel capacity and efficiency [46]. Multiple antennas enhance data transmission and increase the reliability of the system. The number of antennas used in the MIMO system is commonly limited by incorporating multiband antennas that can provide service for different wireless standards [47]. But the design of a MIMO system with multiple antennas of compact size covering multiple bands/wide bandwidth is difficult for portable and handheld devices. The electromagnetic interaction between the antenna elements of a MIMO system where one element receives the energy radiated by the adjacent or nearby element is known as mutual coupling (C). Mathematically, it can be calculated using the following expressions and is usually expressed in dB [48]:

$$C_{mn} = \exp\left(-\frac{2d_{mn}}{\lambda}(\alpha + n\pi) \right),\ m \neq n \tag{1.8}$$

$$C_{mn} = 1 - \frac{1}{N}\sum_{m}\sum_{m \neq n} C_{mn} \tag{1.9}$$

where C_{mn} is the mutual coupling and d_{mn} is the separation between the mth and nth antenna elements. The parameter α is used for controlling the level of coupling, and N is the total number of elements in the MIMO antenna. Another major challenge associated with MIMO antennas is the reduction of the mutual coupling between the individual antenna elements due to the space constraints. The diversity performance of a MIMO system is evaluated using the correlation between the incoming signals at different MIMO ports, denoted by the ECC, which is also related to the mutual coupling. For practical applications, the value of ECC should be less than 0.5.

In this scenario, fractal-shaped antennas can be considered as one of the options that provide solutions for all types of challenges like achieving multiband, broad bandwidth, compactness and improved isolation, ECC, and efficiency in MIMO antenna systems. Recently, many researchers have successfully implemented fractal geometry-based antennas using assorted approaches to meet the demands of modern MIMO systems. In this section, all these techniques are discussed concisely to provide readers with a sound understanding.

1.4.1 Realization techniques

A two-element MIMO antenna utilizing a Koch prefractal edge and a U-shaped slot over a PIFA structure for multistandard use in GSM 1800, the Universal Mobile Telecommunications Service (UMTS), and HiperLAN2 was demonstrated in [49]. The antenna was realized on a 1.57 mm thick RT/Duroid 5880 substrate ($\varepsilon_r = 2.2$) with an air gap of 10 mm. It consists of a simple rectangular patch with its longitudinal resonant edges revamped by the fourth iteration of the Koch fractal geometry to achieve a size miniaturization of 38% compared to the conventional rectangular one. This fractal also supports widening the resonance band ($|S_{11}|$, $|S_{22}| \leq -6$ dB) from 1.82 to 2.19 GHz for GSM 1800 and UMTS applications. The insertion of the U-shaped slot helps in achieving additional resonance band ($|S_{11}|$, $|S_{22}| \leq -6$ dB) between 4.95 and 5.31 GHz, covering the HiperLAN2 application. Measurement indicates that the mutual coupling between elements ($|S_{12}|$) remains below –9 dB and –8 dB, which is suitable for GSM 1800 and UMTS, respectively, whereas it is less than –20 dB at the higher band. Both the PIFA elements were accommodated over a compact size of 100 mm × 45 mm with a finite ground, which perfectly suits implementation in a mobile handset terminal.

A fractal-based hybrid monopole antenna using both the Minkowski island and the Koch geometries at the edge for multiple wireless bands was realized in [50]. For MIMO applications, the two monopoles are placed with an edge-to-edge gap of 0.6 λ_0 at 1.75 GHz. The impedance matching was improved by inserting a T-shaped strip, whereas isolation between the two radiators was enhanced by etching a rectangular slot of appropriate size at the top side of the ground plane. Considering the diversity performance, the antenna has a low ECC value, and the capacity loss value is lower than 0.3 b/s/Hz throughout the usable bands, making it attractive for handheld mobile terminals covering wireless standards like Long-Term Evolution (LTE), Wireless Fidelity (Wi-Fi), Worldwide Interoperability for Microwave Access (WiMAX), and WLAN.

In [51], a four-port UWB MIMO antenna (FUMA) with a frequency rejection characteristic at the WLAN band was presented. The antenna consists of four orthogonally placed octagon-shaped monopole elements miniaturized using Koch fractal geometry at its boundary. Isolation between the elements is achieved by the orthogonal arrangement and further enhanced by including stubs with the ground plane. This ensures that the isolation among different ports of the antenna is better than 17 dB throughout the UWB. The antenna with stub has an IBW ($|S_{11}| \leq -10$ dB) from 2 to 10.6 GHz and a notch band at 5.5 GHz corresponding to the WLAN application. The band rejection characteristic of the antenna is achieved by the inclusion of a C-shaped slot on the top of the monopole. The antenna also has some specialty in terms of its diversity performance along with dimensional compactness, limited to 45 mm × 45 mm. It has a capacity loss and an ECC less than 0.1 b/s/Hz and 0.003, respectively, through the entire band, which signifies a good pulse-preserving characteristic.

Two models of MIMO antenna arrays designed using the Hilbert curve fractal and operating in the ISM band (2.4–2.489 GHz) as well as at higher bands (between 5 and 6 GHz) were demonstrated in [52]. For this antenna, the Hilbert curve was chosen because its higher fractal stages can be utilized for size reduction by occupying the area of the initiator as well as its scaled replicas via a generator rule. Like other fractals, due to the self-similarity of the shape, the Hilbert curve also exhibits multiband behavior. Two prototypes of the antenna based on the first and second stages of the Hilbert space-filling curve (denoted by *Hil_MS1* and *Hil_MS2*, respectively) were designed on an 0.812 mm thick RO3003 substrate ($\varepsilon_r = 3$). Measurement verified that the antenna denoted by *Hil_MS1* has three operational bands of 2.43–2.5 GHz, 4.47–4.68 GHz, and 5.81–5.96 GHz with a correlation coefficient less than 0.1, while the mean effective gain (MEG) does not fall below 5 dB at both the lower and the higher bands. Similarly, the antenna using the second stage of the Hilbert fractal has correlation coefficient values less than 0.1, while the MEG remains under 4 dB for all the spatial directions of the approaching waves at the two operational bands, 2.4–2.62 GHz and 4.53–6.12 GHz. Additionally, the acceptable tolerance of the cross-polarization power ratio (XPR) and the compactness justify the suitability of the antennas for mobile devices using MIMO communication links.

A swastik arm–based two-element MIMO antenna with hepta operating bands covering standards like GSM 900/1800 (0.95–1.02 GHz and 1.73–1.79 GHz), LTE-A (2.68–2.85 GHz), UWB (3.66–3.7 GHz, 4.2–4.4 GHz, and 5.93–6.13 GHz), and WiMAX (5.50–5.65 GHz) was demonstrated in [53]. The antenna geometry was extracted from the hybrid quadratic–Koch island fractal and modified through five development stages to scale down the size to $82 \times 40 \times 0.8$ mm^3. The two radiating elements are separated by a 5 mm space, which helps in maintaining the reflection coefficient value at or below –6 dB and isolation at or above 17 dB without any ancillary structures. As a radiator, the antenna generates CP radiation at 3.66–3.7 GHz and 5.93–6.13 GHz bands and linearly polarized (LP) radiation at other bands. All these characteristics, along with a low ECC value (< 0.05) and a small capacity loss (< 0.5 bits/s/Hz), make the antenna suitable for implementation in 5G communication systems.

A two-port MIMO antenna using a serpentine-shaped fractal radiator with CP radiation characteristics was presented in [54]. The design of the single-element antenna involves three iterations, and in the first iteration, a semicrescent-shaped metallic portion is added to the feed line by removing an elliptical portion of appropriate dimension from a metallic circle with a 2 mm radius. Further, the half size of the semicrescent-shaped portion is connected with 180° phase-shifted semicrescents in the second and third iterations for the generation of two orthogonal modes to achieve CP. The simulated reflection coefficient response shows an IBW of 0.1–3.4 GHz for the final form, but it shifts to 0.1–3 GHz when utilized for the MIMO application. The spatial diversity performance is achieved by arranging two radiators opposite each other on an 0.8 mm thick FR-4 substrate with an area of 8×8 mm^2. The circular polarization performance of the MIMO configuration shows an ARBW from 2.2 to 3.2 GHz. This ARBW is about 24% of the total operational band of 0.1–4.3 GHz, specified for many vital wireless communication standards like LTE, RFID, Wi-Fi, and WiMAX. Moreover, the diversity between left-handed (LH) and right-handed (RH) circular polarization, a wide ARBW, little mutual coupling between the two radiating structures, an acceptable ECC, and compactness make the antenna suitable for implementation in mobile transceivers.

In [55], a decoupling metamaterial (MTM) design based on a fractal electromagnetic bandgap (FEBG) structure was implemented to potentially improve the isolation between the antennas used for transmitting and receiving purposes in a firmly arranged patch array, especially for synthetic aperture radar (SAR) and MIMO systems. The MTM-EMBG structure is a cross-shaped patch with fractal-shaped slots etched in each arm. This fractal geometry consists of four interrelated Y-shaped slots, separated by inverted T-shaped slots. The final form of the antenna was developed by placing the MTM-EMBG structure in a 2×2 antenna array. Investigation affirms that the average isolation between antenna elements 1 and 2, 1 and 3, and 1 and 4 was improved by 17dB, 37dB, and 17dB, respectively, within a frequency band of 8–9.25 GHz (fractional bandwidth of 14.5%) by the implementation of the decoupling structure. The decoupling structure enables the antenna elements to be accommodated with an edge-to-edge separation of 0.5 λ_0 without any degradation in its radiation performance. More importantly, the variation of the antenna gain is limited to between 4 and 7dBi throughout the band, whereas the radiation efficiency varies from 74.22 to 88.71%.

An Amer fractal slot–based four-port MIMO antenna with multifaceted operational modes with bands covering third generation (3G), LTE (2.6 GHz, 3.5 GHz), WLAN (2.4 GHz, 5 GHz), WiMAX (2.5 GHz, 3.5 GHz, 5 GHz), unlicensed ISM (2.4 GHz, 5 GHz), and 5G (5–6 GHz, 27–28 GHz) was demonstrated in [56]. As shown in Figure 1.7(a), the design process starts from the initiator antenna dimensions of 23 $d \times 23$ d and a coplanar ground plane. All the dimensions of the antenna shown in Figure 1.7 are expressed in terms of d, which equals 1 mm. In the initiator antenna, the feed-line width (F) and the gap (g_1) were optimized to be 5 $d \times 1.5$ d and 0.25 d, respectively, for achieving an input impedance of 50 Ω, whereas the separation between the radiator and the coplanar ground plane (g_2) was 0.5 d. In the next step, an Amer fractal slot, as shown in Figure 1.7(b and c), was engraved on the

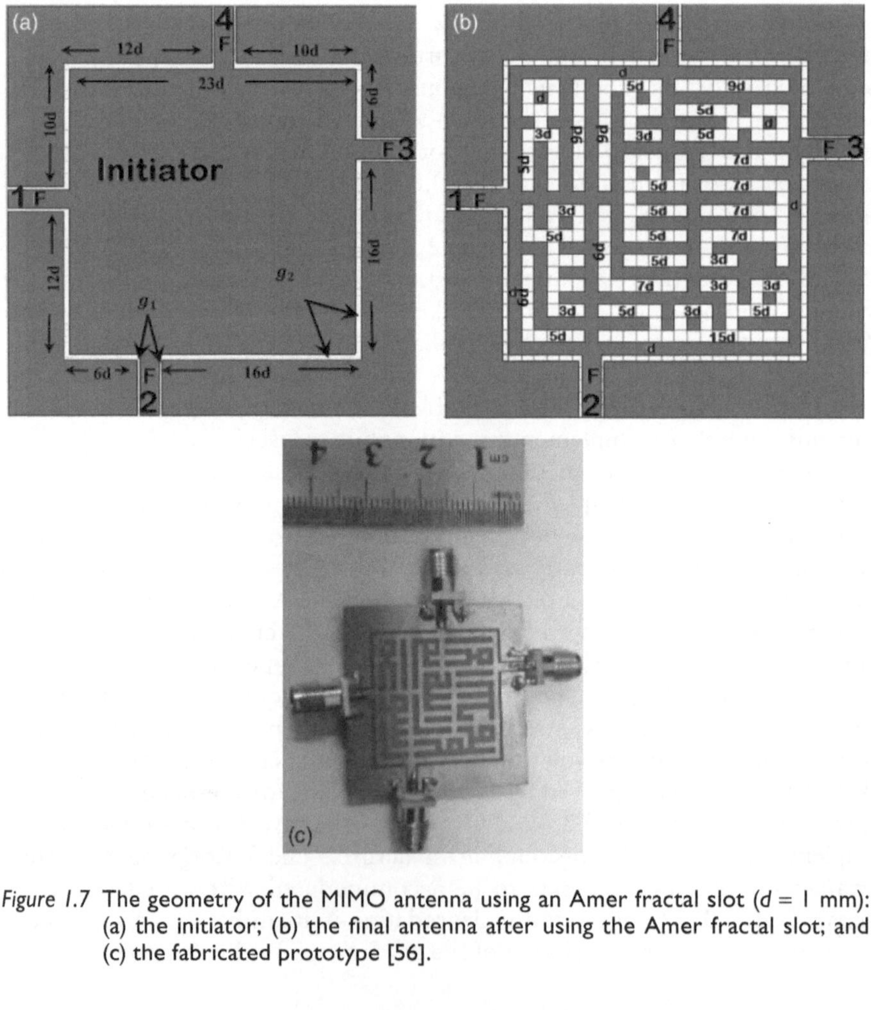

Figure 1.7 The geometry of the MIMO antenna using an Amer fractal slot (d = 1 mm): (a) the initiator; (b) the final antenna after using the Amer fractal slot; and (c) the fabricated prototype [56].

radiator plate of the initiator design to realize the final antenna on a cost-effective FR-4 substrate of 33 × 33 × 0.8 mm^3. From the operational point of view, the antenna has dual operating bands when excited from port 1, port 2, or port 4, whereas for port 3 it shows a wideband characteristic. For the excitation from port 1, the antenna possesses two bands, at 1.5–19.2 GHz and 25–37.2 GHz, while for port 2 the bands shift to 1.4–19 GHz and 20–35.5 GHz. Similarly, for the excitation from port 4, it has two bands, at 1.6–21 GHz and 22–37 GHz, whereas for port 3, the antenna occupies a wide operational bandwidth ranging from 1.4 to 19 GHz. Due to the diverse operational bands, the antenna can be used in different modes like MIMO using two opposite ports within a frequency range of 1.5–15 GHz and as a four-port/element MIMO array at the higher frequency band of 15–30 GHz. Apart from this, the MIMO fractal antenna also generates CP radiation within bandwidths of 4.7–5.8 GHz when excited from port 1, 2.5–2.6 GHz and 5.4–6.5 GHz for port 2, 4–5.9 GHz for port 3, and 5.5–10 GHz for port 4. The ECC values of the antenna between ports 1 and 2, ports 1 and 4, ports 2 and 3, and ports 3 and 4 are less than 0.05, whereas the values for the opposite ports are even smaller, indicating good pattern diversity at all the operating bands.

In [57], a second-iteration Sierpinski carpet FEBG structure was used between two PIFA elements to improve the isolation between them. The measured results show that with FEBG, the coupling levels can be reduced up to 27 dB in the E-plane and 40 dB in the H-plane over the configurations without any isolating structure while maintaining an element spacing less than 0.35 λ_0. The investigation also shows that the resonant frequency of the antenna remains around 2.65 GHz, which is suitable for wireless LTE applications. Further, the antenna possesses a compact dimension of 68 × 40 × 1.6 mm^3, and its ECC is −70 dB, which is smaller than that of antenna elements without FEBG at the operational band, signifying an outstanding diversity characteristic for use in MIMO systems.

In [58], a miniaturized four-port MIMO antenna was designed using a hexagonal molecular fractal geometry–based radiating structure covering a UWB. The first-iteration hexagon molecule fractal at the edge of the antenna helps in obtaining an IBW ($|S_{11}| \leq -10$ dB) of 2.4–10.6 GHz by generating multiple resonances. At the same time, the radiators were accommodated orthogonally to each other within an area of 40 × 40 mm^2 on a 1.6 mm thick RT/Duroid 5880 substrate ($\varepsilon_r = 2.2$) to achieve isolation values higher than 20 dB over the entire range without any decoupling structure. From the measurement, it is observed that the antenna has a notch band around 5.4 GHz for the rejection of the WLAN band, achieved by etching a C-shaped slot on each radiating element. Considering the diversity performance of the antenna, it has an ECC value of 0.02, far below the accepted threshold of 0.5 across the UWB. Further, low-capacity loss (< 0.2 bit/s/Hz) and compactness of the antenna make it suitable for massive MIMO as well as densely packaged mobile devices.

Table 1.2 Comparison of MIMO antennas

Ref.	*No of El.*	*Bandwidth (GHz)*	*ECC*	S_{ij} *(dB)*	*Size (mm³)*	*Comments*
[49]	2	(1.82–2.19) (4.95–5.31)	—	< −6	100 × 45 × 1.57	LP, size is small, mutual coupling is acceptable for MIMO
[50]	2	(1.65–1.9) (2.68–6.25)	0.5	< −10 < −15	100 × 50 × 1.54	LP, capacity loss < 0.3 b/s/Hz throughout the bands
[51]	4	(2–10.6)	0.003	< −17	45 × 45 × 1.6	LP, capacity loss < 0.1 b/s/Hz throughout the band
[52]	2	(2.4–2.489) (5–6)	0.1	< −25	121.8 × 61.45 × 0.812	LP, missed many required bands
[53]	2	LP: (0.95–1.02) (1.73–1.79) (2.68–2.85) (4.2–4.4) (5.5–5.65) CP: (3.66–3.7) (5.93–6.13)	0.05	< −17	82 × 40 × 0.8	Both LP and CP performance, narrow ARBW, lower capacity loss of < 0.5 b/s/Hz
[54]	2	(0.1–4.3)	0.03	< −30	8 × 8 × 0.8	Low gain and efficiency at lower band, missed band of 5–6 GHz
[55]	4	(8–9.25)	—	< −32	120 × 90 × 15	LP, missed many required bands, large size, big edge-to-edge gap
[57]	2	(2.64–2.68)	0.23	< −40	68 × 40 × 1.6	Narrow operating band, LP, missed many quiredbands
[58]	4	(2.4–10.6)	<0.02	< −20	40 × 40 × 1.6	LP, capacity loss < 0.2 bit/s/Hz

Various important characteristics of the above-discussed MIMO antennas are summarized in Table 1.2 for comparison purposes.

1.5 SPECIAL FRACTAL ANTENNAS

In the previous sections, various microstrip antennas designed using fractal geometry for a wide range of wireless applications were discussed. In most cases, these fractal geometries can be mathematically formulated for the generation and implementation of higher-order structures. But recently, many designs have been reported in which the geometries are intricate and difficult to formulate mathematically; yet they are classified as fractals due to the replicative nature of their generation. These special types of fractal antennas are briefly discussed in this section. Some fascinating applications in which fractal antennas have been successfully implemented are also included in this section.

In [59], a modified form of Koch curve was implemented to realize a Fibonacci spiral antenna (FSA) using a 1.5 mm thick RT-Duroid 5870 substrate ($\varepsilon_r = 2.33$, tan $\delta = 0.0012$). The effect of two fractal iterations on antenna size miniaturization and bandwidth was also investigated. The modified Koch curve consists of a semicircular section that acts symmetrically with the quarter-circular sections of the FSA. Unlike the conventional triangular-shaped Koch curve, the modified Koch curve helps to create a smooth flow of surface current due to its curvature. It was found that the FSA using the first iteration occupies a bandwidth of 2.34 –11.50 GHz ($|S_{11}| \leq -5$ dB), while the one using the second iteration operates within the range of 2.76 –10.3 GHz ($|S_{11}| \leq -5$ dB). Dimensional compactness of 50% was achieved by the application of the second iteration of the curve while maintaining a large equivalent arm length corresponding to the lower resonant frequency. However, this size reduction was accomplished at the expense of 27.08% of the bandwidth of the original FSA.

A planar quasi-self-complementary radiator operating in the UWB range from 3.2 to 12 GHz and with a notch band at 5.5 GHz was presented in [60]. In this antenna, impedance matching at a higher band was achieved by implementing the Koch fractal geometry over the hexagonal-shaped radiator, and the ground plane was extended using an arc-shaped portion for impedance matching at the lower frequencies. The band rejection at the above band was obtained by introducing a quadratic fractal slot on the radiator. Additionally, a modified meander-shaped stub was used that generates a resonance at 2.4 GHz for Bluetooth applications. The antenna was designed using an FR-4 substrate of 1.59 mm thickness and occupies a total area of 18.5×39 mm^2.

A planar antipodal Vivaldi antenna using a leaf-shaped fern fractal structure for microwave imaging applications was proposed in [61]. The antenna has a ($|S_{11}| \leq -10$ dB) fractional bandwidth of 175%, ranging from 1.3 to 20 GHz. It was determined that the application of the second-iteration leaf fractal reduces the lower operating frequency of the antenna by 19% compared to the first iteration of the fractal. Along with wide bandwidth, the antenna occupies an area $50.8 \times 62 \times 0.8$ mm^3 (designed on FR-4 substrate, $\varepsilon_r = 4.4$) and has an average gain of 8 dBi with a group delay less than 1 nanosecond.

A wideband antenna operating in the UHF band, ranging from 700 MHz to 4.71 GHz, with compact dimension was presented in [62]. The antenna was developed by modifying both the lateral boundaries of a conventional octagon-shaped monopole and the upper side of the ground plane using Minkowski fractal geometry and concurrently loading the top of the monopole with asymmetric strips. The device miniaturization was accomplished by altering the fractal orientation to make a complementary structure, whereas a triangle-shaped notch was introduced on the ground plane for bandwidth enhancement. The antenna has a small size of 0.28 λ_L × 0.28 λ_L (λ_L = wavelength corresponding to the lower cutoff frequency) and shows an average gain value of 3.93 dBi with good pulse-handling property.

In [63], a compact CP antenna was designed via five iterative steps, and at each step, a modified square patch and half of the portion used in the previous step were combined to form the dual-fractal structure. The measured reflection coefficient performance of the antenna shows that it has operational bands at 2.4–2.65 GHz and 4.8–6.4 GHz for serving Wi-Fi and WiMAX applications. Also, it has a 3 dB ARBW of 35% and 30% at the lower and higher bands, respectively. The antenna has a compact dimension of 18 × 18 × 0.8 mm^3 and can be operated in RH or LH circular polarization mode by altering the input feed point.

In [64], a Koch fractal–based loop rectenna was proposed for radio frequency (RF) energy harvesting. It operates between 1.76 to 1.87 GHz (GSM 1800 band) and consists of three components: a fractal loop antenna, an in-loop ground plane (ILGP), and a rectifier circuit designed using a microstrip configuration. The loop antenna is used for the reception of electromagnetic energy. The implementation of Koch geometry helps in making the antenna compact while enhancing its bandwidth performance. The fractal loop has a length of 1.3 λ at 1.81 GHz and is embedded with an ILGP that improves the input impedance matching performance without altering the dimension of the antenna. A microstrip-based rectifier circuit consisting of a harmonic rejection unit, an impedance matching part, a voltage doubler, and an output filter was also integrated with the antenna to form the rectenna. The rectenna has a compact layout of 45 × 45 × 0.8 mm^3 and an efficiency of 61% at a low-power density of 10μW/cm^2 at 1.8 GHz.

A snowflake fractal antenna formed by four small hexagon-shaped radiators surrounding a larger hexagon and operating at 28 GHz millimeter-wave band was proposed in [65]. The antenna possesses a wide bandwidth ranging from 25.28 to 29.04 GHz (13.43% at 28 GHz) and a 3.12 dBi gain with radiation efficiency greater than 80% throughout the band. Fabricated on a 0.254 mm thick RT/Duroid 5880 substrate (ε_r = 2.2, tan δ = 0.0009), the antenna shows dual-beam radiation throughout the band and occupies an area of 8×5 mm^2. Similarly, a four-element array of the antennas occupies an area of 32×12 mm^2 while gain is enhanced up to 10.12 dBi at 28 GHz.

1.6 SUMMARY AND CONCLUSIONS

In this chapter, a comprehensive analysis of various fractal shapes, as well as their mathematical formulations, advantages, and properties in a broad range of applications, has been presented. The importance of fractal-based planar antennas in communication engineering, especially to cover various wireless standards while helping in the miniaturization process, has been highlighted. The chapter has also demonstrated how the self-similarity and space-filling properties of fractal structures can be incorporated in planar antennas to achieve benefits like compactness, multiband and wideband performance. An extensive analysis on the designs of fractal antennas for UWB, multiband, wideband, RFID, and wearable applications, along with those of some special types of fractal antennas, has been included. Due to the intense and growing research interest in MIMO systems at the current time, a section has been dedicated to the discussion of recently developed fractal-based MIMO antennas used for many contemporary applications.

The application of the fractal concept in the field of antenna engineering began two decades back, and since its inception, it has been applied to various types of conventional antennas, microstrip antennas, and their arrays. Also, many commercial devices are now available in the market that are equipped with fractal antennas, especially the planar type. However, antenna performance is not yet properly correlated with fractal characteristics. The scope for more advanced research on the effect of geometries on antenna performances, which may come up with generic design guidelines for future antenna engineers, is great. There are many existing and upcoming technologies like chipless RFID tag/readers, RF energy harvesting, 5G communication, 5G IoT, massive MIMO, and millimeter-wave communication where the size and number of bands of the antennas play vital roles. Therefore, there is always an opportunity for further research to address the system requirements by designing and exploring new fractal geometry–based MIMO antennas for these applications.

REFERENCES

[1] J. P. Gianvittorio and Y. Rahmat-Samii, "Fractal antennas: A novel antenna miniaturization technique, and applications," *IEEE Antennas and Propagation Magazine*, vol. 44, no. 1, pp. 20–36, 2002.

[2] S. R. Best, "A discussion on the significance of geometry in determining the resonant behavior of fractal and other non-Euclidean wire antennas," *IEEE Antennas and Propagation Magazine*, vol. 45, no. 3, pp. 9–28, 2003.

[3] R. E. Kutter, "Fractal antenna design," B. Sc. honor thesis, University of Dayton, 1996.

[4] D. H. Werner and P. L. Werner, "Frequency-independent features of self-similar fractal antennas," *Radio Science*, vol. 31, no. 6, pp. 1331–1343, 1996.

[5] C. Puente-Baliarda, J. Romeu, R. Pous, and A. Cardama, "On the behavior of the Sierpinski multiband fractal antenna," *IEEE Transactions on Antennas and Propagation*, vol. 46, no. 4, pp. 517–524, 1998.

[6] C. P. Baliarda, J. Romeu, and A. Cardama, "The Koch monopole: A small fractal antenna," *IEEE Transactions on Antennas and Propagation*, vol. 48, no. 11, pp. 1773–1781, 2000.

[7] C. P. Baliarda, C. B. Borau, M. N. Rodero, and J. R. Robert, "An iterative model for fractal antennas: Application to the Sierpinski gasket antenna," *IEEE Transactions on Antennas and Propagation*, vol. 48, no. 5, pp. 713–719, 2000.

[8] J. Pourahmadazar, C. Ghobadi, J. Nourinia, and H. Shirzad, "Multiband ring fractal monopole antenna for mobile devices," *IEEE Antennas and Wireless Propagation Letters*, vol. 9, pp. 863–866, 2010.

[9] A. Azari, "A new super wideband fractal microstrip antenna," *IEEE Transactions on Antennas and Propagation*, vol. 59, no. 5, pp. 1724–1727, 2011.

[10] K. Chen, C. Sim, and J. Row, "A compact monopole antenna for super wideband applications," *IEEE Antennas and Wireless Propagation Letters*, vol. 10, pp. 488–491, 2011.

[11] J. K. Pakkathillam and M. Kanagasabai, "Circularly polarized broadband antenna deploying fractal slot geometry," *IEEE Antennas and Wireless Propagation Letters*, vol. 14, pp. 1286–1289, 2015.

[12] C. Varadhan, J. K. Pakkathillam, M. Kanagasabai, R. Sivasamy, R. Natarajan, and S. K. Palaniswamy, "Triband antenna structures for RFID systems deploying fractal geometry," *IEEE Antennas and Wireless Propagation Letters*, vol. 12, pp. 437–440, 2013.

[13] G. Liu, L. Xu, and Z. Wu, "Dual-band microstrip RFID antenna with tree-like fractal structure," *IEEE Antennas and Wireless Propagation Letters*, vol. 12, pp. 976–978, 2013.

[14] S. R. Best, "On the resonant properties of the Koch fractal and other wire monopole antennas," *IEEE Antennas and Wireless Propagation Letters*, vol. 1, pp. 74–76, 2002.

[15] K. J. Vinoy, J. K. Abraham, and V. K. Varadan, "On the relationship between fractal dimension and the performance of multi-resonant dipole antennas using Koch curves," *IEEE Transactions on Antennas and Propagation*, vol. 51, no. 9, pp. 2296–2303, 2003.

[16] M. Fallahpour and R. Zoughi, "Antenna miniaturization techniques: A review of topology- and material-based methods," *IEEE Antennas and Propagation Magazine*, vol. 60, no. 1, pp. 38–50, 2018.

[17] P. Lande, D. Davis, N. Mascarenhas, F. Fernandes, and A. Kotrashetti, "Design and development of printed Sierpinski carpet, Sierpinski gasket and Koch snowflake fractal antennas for GSM and WLAN applications," in *2015 International Conference on Technologies for Sustainable Development. IEEE*, pp. 1–5, 2015.

[18] S. N. Azemi, A. A. Al-Hadi, R. B. Ahmad, P. J. Soh, and F. Malek, "Multiband fractal planar inverted F antenna (F-PIFA) for mobile phone application," *Progress in Electromagnetics Research*, vol. 14, pp. 127–148, 2009.

[19] K. J. Vinoy, "Fractal shaped antenna elements for wide- and multi-band wireless applications," Ph.D. dissertation, Pennsylvania State University, 2002.

[20] E. Miller and T. Sarkar, "Model-order reduction in electromagnetics using model-based parameter estimation," in *Frontiers in Electromagnetics*, ed. D. H. Werner and R. Mittra, pp. 371–436. Piscataway, NJ: IEEE Press, 2000.

[21] Y. K. Choukiker, "*Investigations on some compact wideband fractal antennas*," Ph.D. dissertation, National Institute of Technology Rourkela, 2013.

[22] D. Chakerian, "Reviewed work: The fractal geometry of nature by Benoit B. Mandelbrot," *College Mathematics Journal*, vol. 15, no. 2, pp. 175–177, 1984.

[23] D. H. Werner and S. Ganguly, "An overview of fractal antenna engineering research," *IEEE Antennas and Propagation Magazine*, vol. 45, no. 1, pp. 38–57, 2003.
[24] N. Cohen, "Fractal antenna applications in wireless telecommunications," in *Professional Program Proceedings: Electronic Industries Forum of New England*. IEEE, pp. 43–49, 1997.
[25] Y. K. Choukiker and S. K. Behera, "Modified Sierpinski square fractal antenna covering ultra-wide band application with band notch characteristics," *IET Microwaves, Antennas and Propagation*, vol. 8, no. 7, pp. 506–512, 2014.
[26] S. Tripathi, A. Mohan, and S. Yadav, "Hexagonal fractal ultra-wideband antenna using Koch geometry with bandwidth enhancement," *IET Microwaves, Antennas and Propagation*, vol. 8, no. 15, pp. 1445–1450, 2014.
[27] S. Singhal and A. K. Singh, "CPW-fed hexagonal Sierpinski super wideband fractal antenna," *IET Microwaves, Antennas and Propagation*, vol. 10, no. 15, pp. 1701–1707, 2016.
[28] N. K. Darimireddy, R. R. Reddy, and A. M. Prasad, "A miniaturized hexagonal-triangular fractal antenna for wide-band applications [antenna applications corner]," *IEEE Antennas and Propagation Magazine*, vol. 60, no. 2, pp. 104–110, 2018.
[29] J. Liang, C. C. Chiau, X. Chen, and C. G. Parini, "Study of a printed circular disc monopole antenna for UWB systems," *IEEE Transactions on Antennas and Propagation*, vol. 53, no. 11, pp. 3500–3504, 2005.
[30] D. Schaubert, E. Kollberg, T. Korzeniowski, T. Thungren, J. Johansson, and K. Yngvesson, "Endfire tapered slot antennas on dielectric substrates," *IEEE Transactions on Antennas and Propagation*, vol. 33, no. 12, pp. 1392–1400, 1985.
[31] P. Hall, "New wideband microstrip antenna using log-periodic technique," *Electronics Letters*, vol. 16, no. 4, pp. 127–128, 1980.
[32] A. Amini, H. Oraizi, and M. A. Chaychi zadeh, "Miniaturized UWB log-periodic square fractal antenna," *IEEE Antennas and Wireless Propagation Letters*, vol. 14, pp. 1322–1325, 2015.
[33] F. Viani, M. Salucci, F. Robol, G. Oliveri, and A. Massa, "Design of a UHF RFID/GPS fractal antenna for logistics management," *Journal of Electromagnetic Waves and Applications*, vol. 26, no. 4, pp. 480–492, 2012.
[34] F. Viani, M. Salucci, F. Robol, and A. Massa, "Multiband fractal Zigbee/WLAN antenna for ubiquitous wireless environments," *Journal of Electromagnetic Waves and Applications*, vol. 26, no. 11–12, pp. 1554–1562, 2012.
[35] W. C. Weng and C.-L. Hung, "An H-fractal antenna for multiband applications," *IEEE Antennas and Wireless Propagation Letters*, vol. 13, pp. 1705–1708, 2014.
[36] A. K. M. M., A. Patnaik, and C. G. Christodoulou, "Design and testing of a multifrequency antenna with a reconfigurable feed," *IEEE Antennas and Wireless Propagation Letters*, vol. 13, pp. 730–733, 2014.
[37] S. Sankaralingam and B. Gupta, "Determination of dielectric constant of fabric materials and their use as substrates for design and development of antennas for wearable applications," *IEEE Transactions on Instrumentation and Measurement*, vol. 59, no. 12, pp. 3122–3130, 2010.

[38] C.-P. Lin, C.-H. Chang, Y. T. Cheng, and C. F. Jou, "Development of a flexible SU-8/PDMS-based antenna," *IEEE Antennas and wireless propagation letters*, vol. 10, pp. 1108–1111, 2011.
[39] S. Velan, E. F. Sundarsingh, M. Kanagasabai, A. K. Sarma, C. Raviteja, R. Sivasamy, and J. K. Pakkathillam, "Dual-band EBG integrated monopole antenna deploying fractal geometry for wearable applications," *IEEE Antennas and Wireless Propagation Letters*, vol. 14, pp. 249–252, 2014.
[40] A. Arif, M. Zubair, M. Ali, M. U. Khan, and M. Q. Mehmood, "A compact, low-profile fractal antenna for wearable on-body WBAN applications," *IEEE Antennas and Wireless Propagation Letters*, vol. 18, no. 5, pp. 981–985, 2019.
[41] F. Mokhtari-Koushyar, P. M. Grubb, M. Y. Chen, and R. T. Chen, "A miniaturized tree-shaped fractal antenna printed on a flexible substrate: A lightweight and low-profile candidate with a small footprint for spaceborne and wearable applications," *IEEE Antennas and Propagation Magazine*, vol. 61, no. 3, pp. 60–66, 2019.
[42] N. C. Karmakar, *Handbook of Smart Antennas for RFID Systems*. Hoboken, NJ: Wiley, 2011.
[43] A. Farswan, A. K. Gautam, B. K. Kanaujia, and K. Rambabu, "Design of Koch fractal circularly polarized antenna for handheld UHF RFID reader applications," *IEEE Transactions on Antennas and Propagation*, vol. 64, no. 2, pp. 771–775, 2015.
[44] T. K. Das, B. Dwivedy, D. Behera, S. K. Behera, and N. C. Karmakar, "Design and modelling of a compact circularly polarized antenna for RFID applications," *AEU—International Journal of Electronics and Communications*, vol. 123, pp. 1–9, 2020.
[45] C. Raviteja, C. Varadhan, M. Kanagasabai, A. K. Sarma, and S. Velan, "A fractal-based circularly polarized UHF RFID reader antenna," *IEEE Antennas and Wireless Propagation Letters*, vol. 13, pp. 499–502, 2014.
[46] S. Kumar, A. S. Dixit, R. R. Malekar, H. D. Raut, and L. K. Shevada, "Fifth generation antennas: A comprehensive review of design and performance enhancement techniques," *IEEE Access*, vol. 8, pp. 163568–163593, 2020.
[47] R. Khan, A. A. Al-Hadi, P. J. Soh, M. R. Kamarudin, M. T. Ali, and Owais, "User influence on mobile terminal antennas: A review of challenges and potential solution for 5G antennas," *IEEE Access*, vol. 6, pp. 77695–77715, 2018.
[48] I. Nadeem and D. Choi, "Study on mutual coupling reduction technique for MIMO antennas," *IEEE Access*, vol. 7, pp. 563–586, 2019.
[49] J. Guterman, A, A. Moreira, and C. Peixeiro, "Microstrip fractal antennas for multistandard terminals," *IEEE Antennas and Wireless Propagation Letters*, vol. 3, pp. 351–354, 2005.
[50] Y. K. Choukiker, S. K. Sharma, and S. K. Behera, "Hybrid fractal shape planar monopole antenna covering multiband wireless communications with MIMO implementation for handheld mobile devices," *IEEE Transactions on Antennas and Propagation*, vol. 62, no. 3, pp. 1483–1488, 2014.
[51] S. Tripathi, A. Mohan, and S. Yadav, "A compact Koch fractal UWB MIMO antenna with WLAN band-rejection," *IEEE Antennas and Wireless Propagation Letters*, vol. 14, pp. 1565–1568, 2015.

[52] A. Peristerianos, A. Theopoulos, A. G. Koutinos, T. Kaifas, and K. Siakavara, "Dual-band fractal semi-printed element antenna arrays for MIMO applications," *IEEE Antennas and Wireless Propagation Letters*, vol. 15, pp. 730–733, 2016.
[53] S. Rajkumar, N. V. Sivaraman, S. Murali, and K. T. Selvan, "Heptaband swastik arm antenna for MIMO applications," *IET Microwaves, Antennas and Propagation*, vol. 11, no. 9, pp. 1255–1261, 2017.
[54] A. T. Abed, "Highly compact size serpentine-shaped multiple-input-multiple-output fractal antenna with CP diversity," *IET Microwaves, Antennas and Propagation*, vol. 12, no. 4, pp. 636–640, 2018.
[55] M. Alibakhshikenari, B. S. Virdee, C. H. See, R. Abd-Alhameed, A. Hussein Ali, F. Falcone, and E. Limiti, "Study on isolation improvement between closely-packed patch antenna arrays based on fractal meta-material electromagnetic bandgap structures," *IET Microwaves, Antennas and Propagation*, vol. 12, no. 14, pp. 2241–2247, 2018.
[56] A. T. Abed and A. M. Jawad, "Compact size MIMO Amer fractal slot antenna for 3G, LTE (4G), WLAN, WiMAX, ISM and 5G communications," *IEEE Access*, vol. 7, pp. 125542–125551, 2019.
[57] A. H. Radhi, R. Nilavalan, Y. Wang, H. Al-Raweshidy, A. A. Eltokhy, and N. A. Aziz, "Mutual coupling reduction with a novel fractal electromagnetic bandgap structure," *IET Microwaves, Antennas and Propagation*, vol. 13, no. 2, pp. 134–141, 2019.
[58] S. Rajkumar, A. A. Amala, and K. T. Selvan, "Isolation improvement of UWB MIMO antenna utilising molecule fractal structure," *Electronics Letters*, vol. 55, no. 10, pp. 576–579, 2019.
[59] C. Sharma and D. K. Vishwakarma, "Miniaturization of spiral antenna based on Fibonacci sequence using modified Koch curve," *IEEE Antennas and Wireless Propagation Letters*, vol. 16, pp. 932–935, 2017.
[60] A. Gorai, M. Pal, and R. Ghatak, "A compact fractal-shaped antenna for ultrawideband and Bluetooth wireless systems with WLAN rejection functionality," *IEEE Antennas and Wireless Propagation Letters*, vol. 16, pp. 2163–2166, 2017.
[61] B. Biswas, R. Ghatak, and D. R. Poddar, "A fern fractal leaf inspired wideband antipodal Vivaldi antenna for microwave imaging system," *IEEE Transactions on Antennas and Propagation*, vol. 65, no. 11, pp. 6126–6129, 2017.
[62] F. Wang, F. Bin, Q. Sun, J. Fan, and H. Ye, "A compact UHF antenna based on complementary fractal technique," *IEEE Access*, vol. 5, pp. 21118–21125, 2017.
[63] A. T. Abed, M. S. J. Singh, and M. T. Islam, "Compact fractal antenna circularly polarized radiation for Wi-Fi and WiMAX communications," *IET Microwaves, Antennas and Propagation*, vol. 12, no. 14, pp. 2218–2224, 2018.
[64] M. Zeng, A. S. Andrenko, X. Liu, Z. Li, and H. Tan, "A compact fractal loop rectenna for RF energy harvesting," *IEEE Antennas and Wireless Propagation Letters*, vol. 16, pp. 2424–2427, 2017.
[65] H. Ullah and F. A. Tahir, "A novel snowflake fractal antenna for dual-beam applications in 28 GHz band," *IEEE Access*, vol. 8, pp. 19873–19879, 2020.

Chapter 2

Multiband hybrid fractal antennas for smart networks

Yadwinder Kumar, Mandeep Singh, and Manjit Kaur

CONTENTS

2.1 INTRODUCTION

It is hard to imagine today's world without communication; similarly, the survival of wireless communication can't be envisioned without antennas. The heart of all wireless communication systems is the antenna. It changes electric signals into electromagnetic radio waves that travel into free space. It also exhibits a property of reciprocity, which suggests that it will maintain the same characteristics whether it is a transmitting or a receiving antenna. The need for concise, handy, and multiband antennas has gradually grown with time. Enormous advances in antenna design, structure, and materials are the result of years of investigation and development in

DOI: 10.1201/9781003290230-2

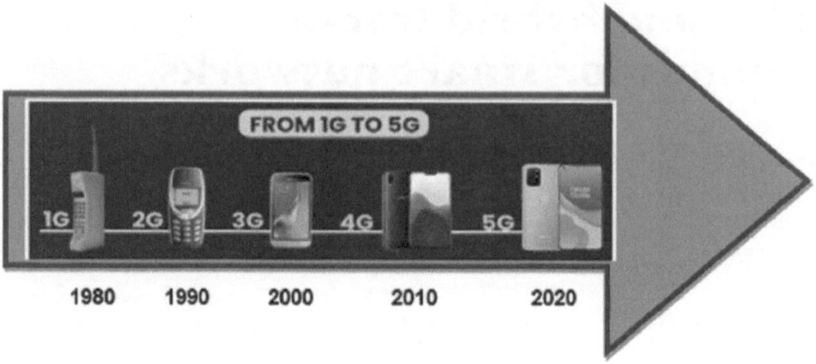

Figure 2.1 Mobile communication journey from 1G to 5G.

this domain. This process of advancement is still continuing in an effort to develop more optimized and beneficial prototypes.

Figure 2.1 shows the journey of mobile communication from an era of bulky but portable handsets to today's communication devices. This journey from first generation (1G) to fifth generation (5G) has been a gradual success story. External wire antennas were slowly replaced by integrated antennas, which were so small and compact that they hardly occupied any significant space inside the device. This was made possible by a significant development in the field of planar antennas. Integrated planar microstrip multiband antennas are now broadly utilized in different recurrence groups and technologies like the Global System for Mobile Communication (GSM), the Global Positioning System (GPS), Bluetooth, Wireless Fidelity (Wi-Fi), Near-Field Communication, Long-Term Evolution and Worldwide Interoperability for Microwave Access [1].

Compact optimized antennas capable of operating at multiband frequencies are now a major requirement of wireless communications systems. The study of fractal antennas has developed in order to satisfy this requirement. In 1953, Mandelbrot introduced fractal geometries [1]. Later on, fractal geometries were explored and validated for use in wireless communication systems. Due to their inherit characteristics like self-similarity and space filling, fractal antennas can exhibit compactness and multiband behavior [2]. As fractal geometries are complex in nature, it is not necessary that they resonate at user-defined frequencies [3–5]. The divide between the working frequencies of a fractal structure further fueled interest for their steering and calibration considering annoyance of the configuration framework [6, 7].

It is a challenging task for designers to use fractals as radiating structures that resonate at user-defined frequencies. In order to achieve this, a time-consuming trial-and-error method has been widely adopted by researchers, in which calculated design parameters are adjusted or modified until the desired outcome is obtained [8, 9]. Such experiments have been done manually and repeated numerous times till the target is achieved. In order to reduce the large amount of time consumed by the trial-and-error method, various optimization methods like genetic algorithm, artificial neural network, ant colony and particle swarm optimization can also be employed [10–12].

Antennas can be grouped according to their build style, qualities and applications. They can be categorized as linear wire, reflector, physical aperture, microstrip patch, slot and lens, and, finally, smart antennas like multiple-input multiple-output (MIMO) antennas [13–15].

As the demand for multiple wireless standards has arisen, multiple antennas have needed to be replaced by a single multiband antenna, and fractal antennas have emerged as the most suitable choice for these requirements and applications. Fractal antennas are low profile, low cost, and planar like microstrip patch antennas. They can be fabricated on different types of substrate materials having distinctive widths as demanded by the application.

Fractal antennas have the capacity to satisfy every one of these needs because of their self-similarity and space-filling properties and their capacity to be scaled down [2]. Fractal calculations were first proposed by Mandelbrot in 1953 [16] and afterward were generally embraced in planning receiving antennas for multiband/wideband correspondences. Fractal structures have unique properties obtained by periodic activity that expand the viable electrical length in a constrained volume. They also help in accomplishing multiband behavior and compactness [17]. Extensively used fractals like Koch, Sierpinski, Cantor set, and tree have been effectively utilized to deliver multiband antennas for wireless applications [18].

2.2 CREATING FRACTALS USING THE ITERATED FUNCTION SYSTEM METHOD

The iterated function system (IFS) is an amazingly valuable and flexible scientific strategy for producing a wide assortment of fractal geometries by characterizing change, scaling, and the turn of initiator layout [19]. These iterated frameworks are based on the arrangement of relative changes. The composing matrix representation of lone sections produces changes to get portions of the generator. A change to get a fragment of the generator is given by

$$W\begin{pmatrix} x \\ y \end{pmatrix} = \begin{pmatrix} a & b \\ c & d \end{pmatrix}\begin{pmatrix} x \\ y \end{pmatrix} + \begin{pmatrix} e \\ f \end{pmatrix} \tag{2.1}$$

or, alternatively, by

$$w(x, y) = (ax + by + e,\ cx + dy + f) \tag{2.2}$$

where a and f denote real numbers. These are used to steer the rotation, scaling, and linear shift or relocation, as shown in Figure 2.2 [20].

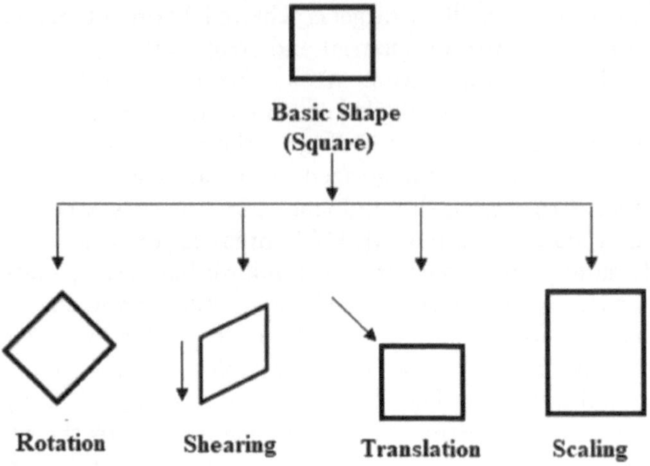

Figure 2.2 Affine transformation showing rotation, shearing, translation, and scaling.

Assume A is the initial starting geometry. With the application of a set of transformations to A [4], it could be denoted as

$$W(A) = \bigcup_{n=1}^{N} Wn(A) \tag{2.3}$$

IFS also provides a great deal of flexibility, allowing the conventional shapes to be modified or customized to generate new, unique, and novel fractal structures.

2.2.1 Modified Koch curve fractal

The conventional Koch curve has an edge length of 1/3, which can be modified, Figure 2.3 shows a modified version of a Koch curve fractal with an edge length of 0.5 at the first, second, and third iterations. These new fractal shapes can be used to generate radiating structures with the help of electromagnetic simulation software [21]. Planar fractal antennas like microstrip antennas can be generated using such novel shapes. Such fractal shapes have a large number of edges, corners, and fragments, which enable them to get different resonances [18]. As the number of segments increases, the total effective length also increases in the case of line fractals. Table 2.1 represents the IFS of the modified Koch curve in rotation form.

2.2.2 Modified Sierpinski gasket fractal

Similarly, other popular conventional fractal shapes like the Sierpinski gasket can be further modified to generate various fractal and prefractal

Figure 2.3 (a) Standard Koch curve; (b) modified Koch curve having an edge length of 0.5 first iteration; (c) second iteration; and (d) third iteration.

shapes, and various other possibilities can be explored [22]. Such promising shapes as radiating shapes can be used in microstrip antenna structure arrangements. These types of structures are among the radiating structures most preferred by researchers in designing smart MIMO antennas [23].

Figure 2.4 shows the modified Sierpinski triangle fractal with customized positions of factors a, b, and c; further, it can be generated with many iterations. Table 2.2 shows the IFS functions of the modified Sierpinski fractal in rotation form.

Table 2.1 IFS functions of the modified Koch fractal having an edge length of 0.5

	Scale		*Rotation*		*Relocation*	
	a	*b*	*c*	*d*	*e*	*f*
1	0.250	0.250	0.00	0.00	0.00	0.00
2	0.500	0.500	60.00	60.00	0.250	0.00
3	0.500	0.500	–60.0	–60.0	0.500	0.433
4	0.250	0.250	0.00	0.00	0.750	0.00

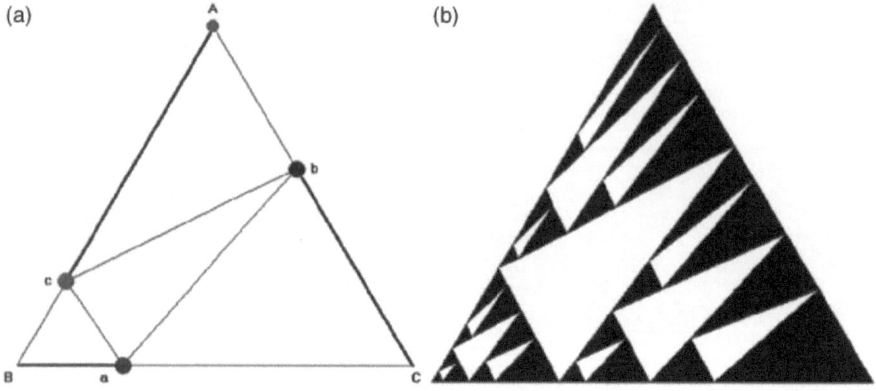

Figure 2.4 Modified Sierpinski triangle fractal: (a) customized generator shape at iteration 1; (b) final shape with iteration 3.

2.2.3 Modified Pythagorean tree fractal

The Pythagorean tree fractal is also a very widely used and popular fractal shape for designing multiband planar antennas [24]. This fractal structure in its conventional shape has an angle of 45° between the upper square (left arm) and the base square, as shown in Figure 2.5(a). As discussed earlier, fractal shapes have great flexibility in terms of the modification of their basic shapes. In this fractal, we can change the angle from 45° to 60°, and the resultant shape is shown in Figure 2.5(b). Table 2.3 shows the IFS functions of the modified Pythagorean tree fractal with a 60° angle.

2.3 HYBRID FRACTAL ANTENNAS

2.3.1 Koch-meander hybrid fractal antenna

Hybrid fractal antennas can be laid out either by coalescing, joining, merging, or superimposing two different fractal shapes or by merging nonfractal

Table 2.2 IFS functions of the modified Sierpinski gasket fractal

	Scale		*Rotation*		*Relocation*	
	a	*b*	*c*	*d*	*e*	*f*
1	0.289	−0.301	0.00	−1.36	0.00	0.00
2	0.711	0.628	0.00	−4.47	0.289	0.00
3	0.606	0.544	27.71	−9.93	0.151	0.261

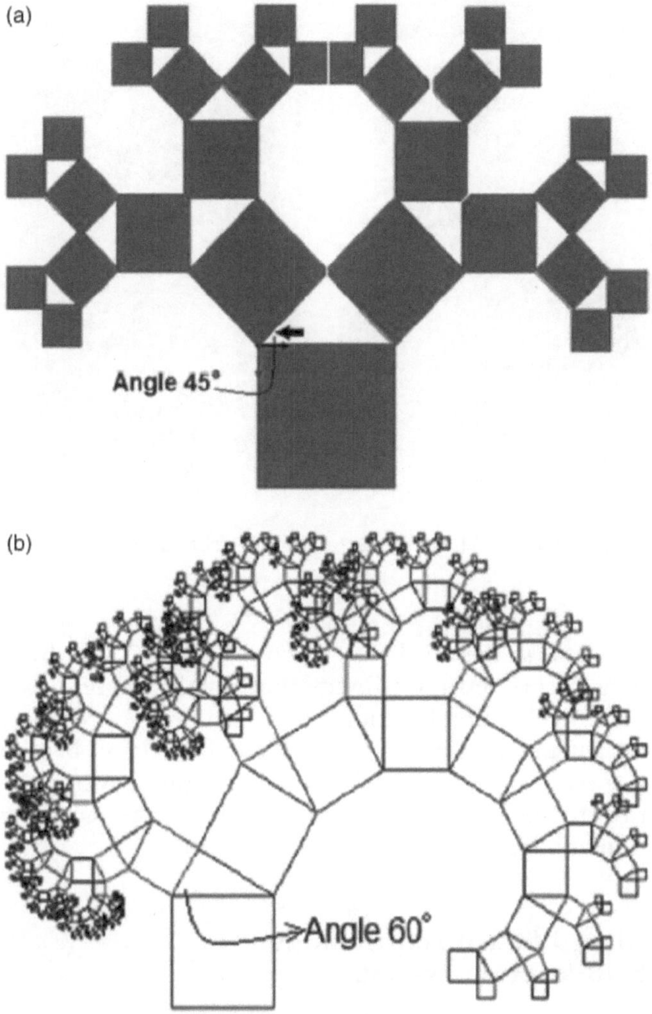

Figure 2.5 (a) Conventional Pythagorean tree fractal with a 45° angle; (b) modified Pythagorean tree fractal with a 60° angle.

Table 2.3 IFS functions of the modified Pythagorean tree fractal with a 60° angle

	Scale		*Rotation*		*Relocation*	
	a	*b*	*c*	*d*	e	*f*
1	0.500	0.500	60.00	60.00	0.00	1.00
2	0.866	0.866	−30.0	−30.0	0.250	1.433
3	1.00	1.00	0.00	0.00	0.00	0.00

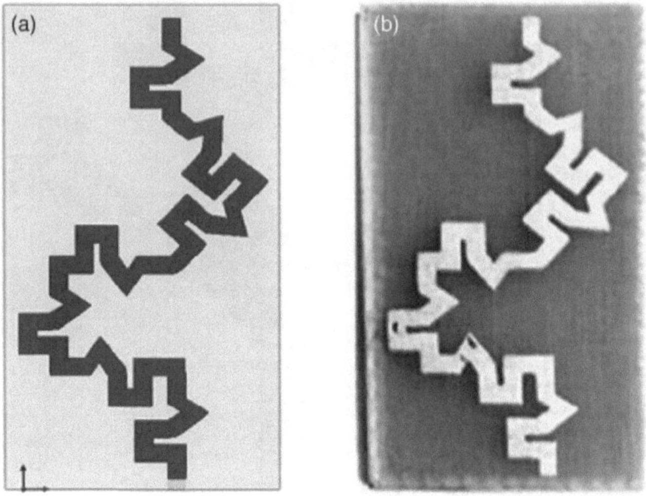

Figure 2.6 (a) Koch-Minkowski hybrid fractal [26]; (b) fabricated antenna prototype.

shapes with fractal shapes [25]. As shown in Figure 2.6(a), by adding the popular Koch fractal and the Minkowski fractal, a hybrid fractal shape is derived [26]. Such a combination turns out to be extremely useful for designing an antenna structure, as it exhibits heptaband behavior; i.e., it resonates at seven different useful frequencies. Figure 2.6(b) shows the fabricated hardware antenna on an FR4 substrate. Such a compact antenna can be used in handheld portable wireless devices.

2.3.2 Modified Sierpinski-meander hybrid fractal antenna

Another hybrid fractal, shown in Figure 2.7(a), has been designed by combining a modified Sierpinski fractal and a meander antenna [27]. This structure exhibits heptaband behavior by resonating at seven different useful frequencies, including Bluetooth; Wireless Local Area Network (WLAN); Wi-Fi; Industrial, Scientific and Medical; Radio Frequency Identification; and Long-Term Evolution. The hardware antenna prototype fabricated on an FR4 substrate is shown in Figure 2.7(b).

2.3.3 Koch–Sierpinski gasket–Sierpinski carpet hybrid fractal antenna

Another hybrid fractal layout is formed by combining the Koch curve, Sierpinski gasket, and Sierpinski carpet. As is clear from Figure 2.8, which

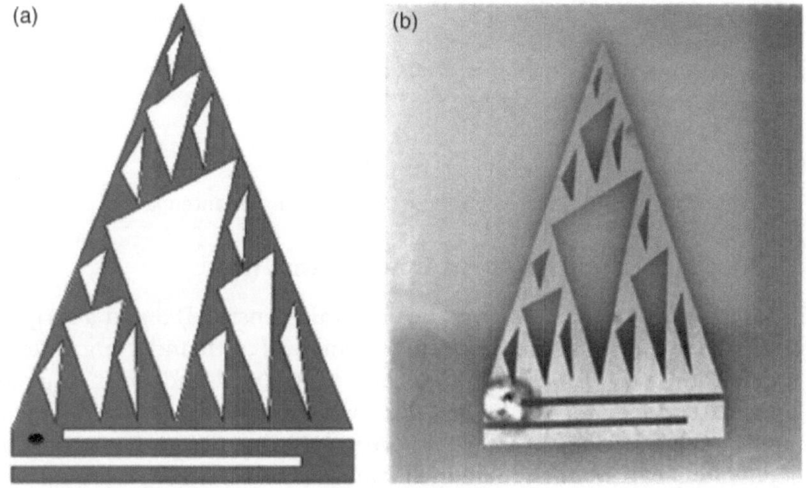

Figure 2.7 (a) Modified Sierpinski-meander hybrid fractal antenna; (b) fabricated antenna prototype developed on an FR4 substrate [27].

shows the hardware antenna prototype fabricated on an FR4 substrate, the Koch curve in its conventional form is added on the four boundaries of the Sierpinski carpet. Both the popular fractal shapes are joined to form a hybrid fractal layout without modifying their basic shapes. The resulting antenna resonates on seven useful frequencies [28].

Figure 2.8 Koch-Sierpinski hybrid fractal antenna prototype developed on an FR4 substrate.

Figure 2.9 Modified Koch fractal curve and modified meander antenna.

2.3.4 Koch-meander hybrid fractal antenna

A proposed hybrid fractal antenna layout drafted using IFS and a scripting method is depicted in Figure 2.9 [29]. It combines a Meander line antenna (with indention angle $\theta = 90°$) and a Koch curve (with indention angle $\theta = 60°$). The Meander and Koch fractal curves are broadly utilized in wireless applications like GPS, GSM, Universal Mobile Telecommunications Service, Wi-Fi, WLAN, and Bluetooth and in smart MIMO antennas [30].

This fractal geometry has been obtained by using *W*, as mentioned in Equation (2.1). The changes to get the fragments of the initiator of the proposed layout are shown in Table 2.4.

The generator has been obtained by combining all of the quantities shown in Table 2.4 [1].

$$W(A) = W_1(A)\ldots\ldots\ \mathrm{U}W_7(A) \tag{2.4}$$

This procedure can be replicated for higher iterations. Every fragment with which to begin the cycle (generator) is found in 1/5 of the reference structure. There are seven such portions. Fractal likeness measurements can be calculated using the following expression [6]:

$$\mathrm{D} = \frac{\log(n)}{\log(r)} = \frac{\log 7}{\log 5} = 1.20906 \tag{2.5}$$

The suggested layout of the hybrid fractal antenna shown in Figure 2.10 is derived from the second iteration of the reference structure created by combining the modified Koch and modified Meander fractal curves.

Table 2.4 IFS quantities used for the proposed hybrid fractal layout

w	a	b	c	d	e	f
1	0.2	0	0	0.2	0	0
2	0.2	–0.25	0.25	0.2	0.2	0
3	0	–0.25	0.25	0	0.4	0.25
4	0.2	0	0	0.2	0.4	0.25
5	0	0.25	–0.25	0	0.6	0.5
6	1/5	1/4	–1/4	1/5	3/5	1/4
7	1/5	0	0	1/5	4/5	0

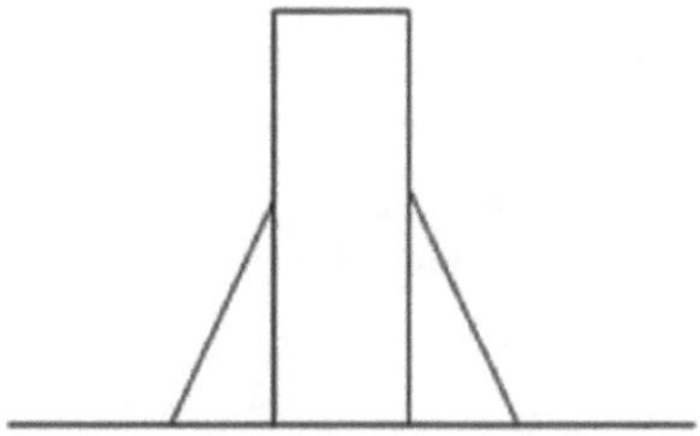

Figure 2.10 Resultant layout of the proposed hybrid fractal initiator.

A MATLAB program has been developed using the IFS to generate VBScript (.vbs file) for the proposed hybrid fractal geometry. The VBScript file has then been imported to an electromagnetic field solver using a scripting method [27, 31].

2.4 RESULTS AND DISCUSSION

The layout of the suggested hybrid fractal antenna was developed using methodologies discussed above, and the final structure, shown in Figure 2.11, was obtained by using the initiator structure discussed in the previous section. The radiating structure has been designed in the form of

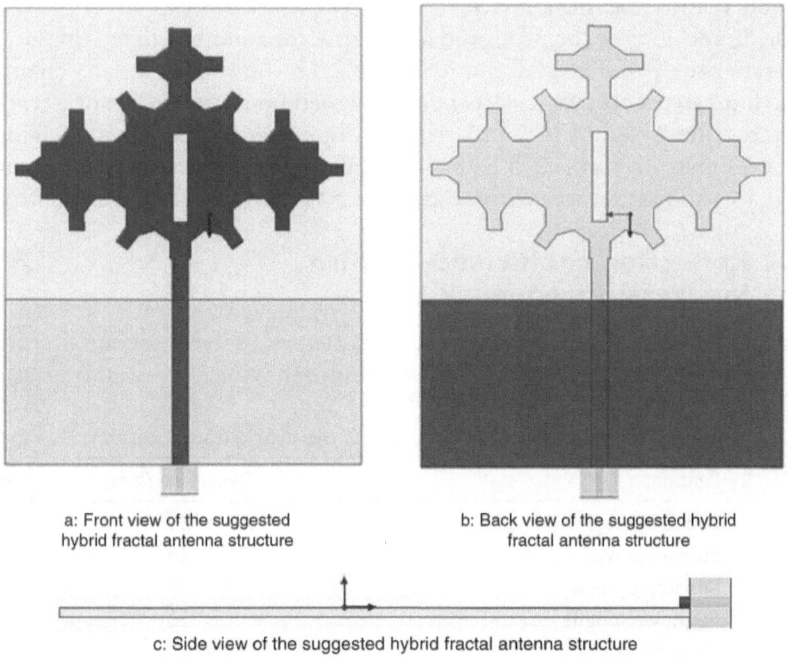

Figure 2.11 Hybrid fractal antenna schematic for the second iteration: (a) front view; (b) back view; and (c) side view.

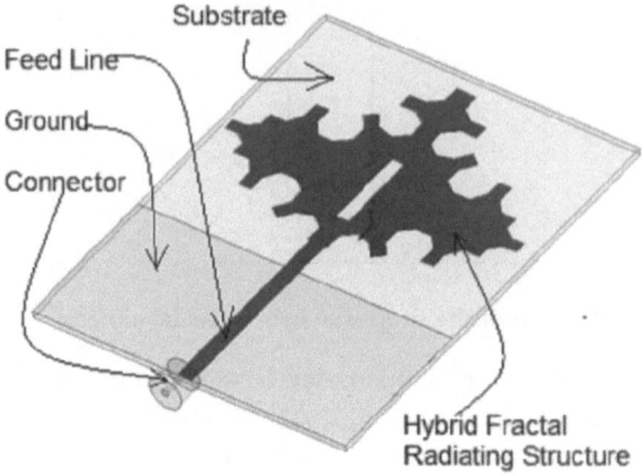

Figure 2.12 Hybrid fractal antenna schematic: detailed tilted view.

a dipole [32] antenna with some perturbations on the reference geometry in order to get the desired results. Figure 2.11(a) depicts the front view, Figure 2.11(b) depicts the back view (with some level of transparency), and the colored portions show the copper area. This copper area can be either deposited or etched on the FR4 substrate. Figure 2.11(c) shows the side view; it has a very slim and compact profile, so it can be easily placed in any portable communication device.

The dimensions of the proposed hybrid fractal antenna etched on the FR4 substrate are 53 × 38 × 1.6 mm^3. Figure 2.12 shows a detailed schematic of the final structure, which has been designed up to two levels of iteration. The structure is fed by a microstrip feed line that lies on the same side of the FR4 substrate. Table 2.5 lists all the geometrical parameters of the proposed antenna structure, with all dimensions represented in millimeters.

2.4.1 Reflection coefficient—return loss versus frequency graph

Figure 2.13 shows the return loss parameters for the second iteration. The proposed antenna exhibits pentaband behavior by resonating at five

Table 2.5 Geometrical parameters of the hybrid fractal antenna

Parameter	*Dimension (mm)*
Substrate length	53
Substrate width	38
Substrate thickness	1.6
Feed-line length	28
Ground length	22.5
Ground width	38
Slot length	10
Slot width	2

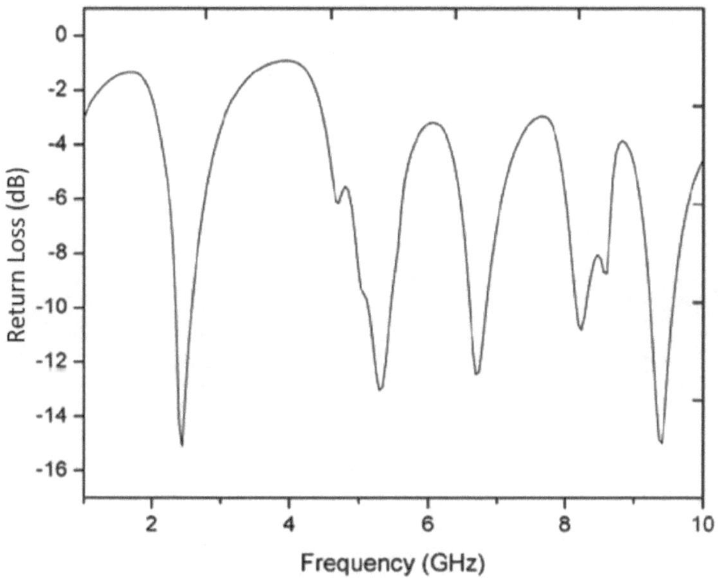

Figure 2.13 Return loss graph of the hybrid fractal antenna for the second iteration.

different frequencies — 2.441, 5.341, 6.743, 8.236, and 9.421 GHz—with acceptable values of return loss.

2.4.2 Voltage standing wave ratio

According to the maximum power transfer theorem, maximum power can be delivered to an antenna if the output impedance of the transmission line is equal to the input impedance of the antenna. This degree of impedance matching is represented by a parameter called the voltage standing wave ratio (VSWR) [13]. The mismatch is usually represented by the reflection coefficient, which is a measure of the power reflected back by the antenna. Figure 2.14 shows a graph of the VSWR versus the frequency. For each resonating frequency, the VSWR value obtained is clearly less than 2. The ideal range lies between 1 and 2.

The kind of material utilized for the substrate plays an exceptionally crucial part in boosting the execution characteristics of the antenna structure [33]. The choice of material is the first and most important step in the process used to obtain the required results. In this detailed analysis, we selected four very commonly used substrate materials: Teflon, RT/Duriod 5880, quartz glass, and FR4, which have dielectric constants of $\epsilon_r = 2.1$, $\epsilon_r = 2.2$, $\epsilon_r = 3.1$, and $\epsilon_r = 4.4$, respectively. A detailed analysis of these materials and their impact on different parameters of the antenna is given in Table 2.6. Valuable and satisfactory results were achieved only for the FR4 substrate.

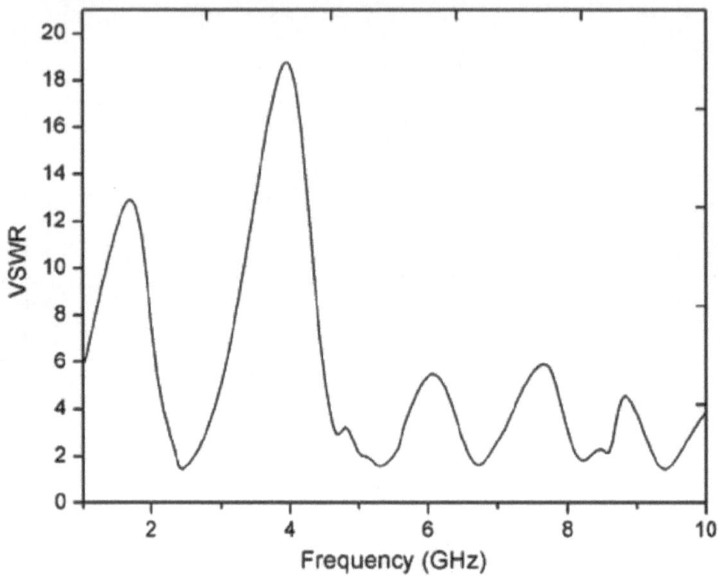

Figure 2.14 VSWR versus frequency graph for the hybrid fractal antenna for the second iteration.

Table 2.6 Behavioral analysis of proposed antenna for different substrates materials

Dielectric constant	*Resonant frequency (GHz)*	*Return loss (dB)*	*Input impedance (ohm)*	*VSWR*
2.1 (Teflon)	7.74	−17.73	61.82	1.31
	6.72	−12.42	51.23	1.59
	8.34	−27.85	49.61	1.26
2.2 (RT/Duriod)	4.32	−15.82	61.87	1.41
	3.61	−13.76	39.57	1.34
	9.75	−12.73	41.46	1.47
	9.34	−26.39	56.98	1.12
3.1 (quartz glass)	5.34	−15.23	42.23	3.74
	6.71	−28.62	58.32	2.82
	6.41	−14.67	51.23	3.12
4.4 (FR4)	2.441	−15.0	51.71	1.426
	5.3417	−12.94	54.36	1.583
	6.743	−12.36	61.05	1.627
	8.236	−10.82	50.90	1.807
	9.421	−14.97	58.32	1.441

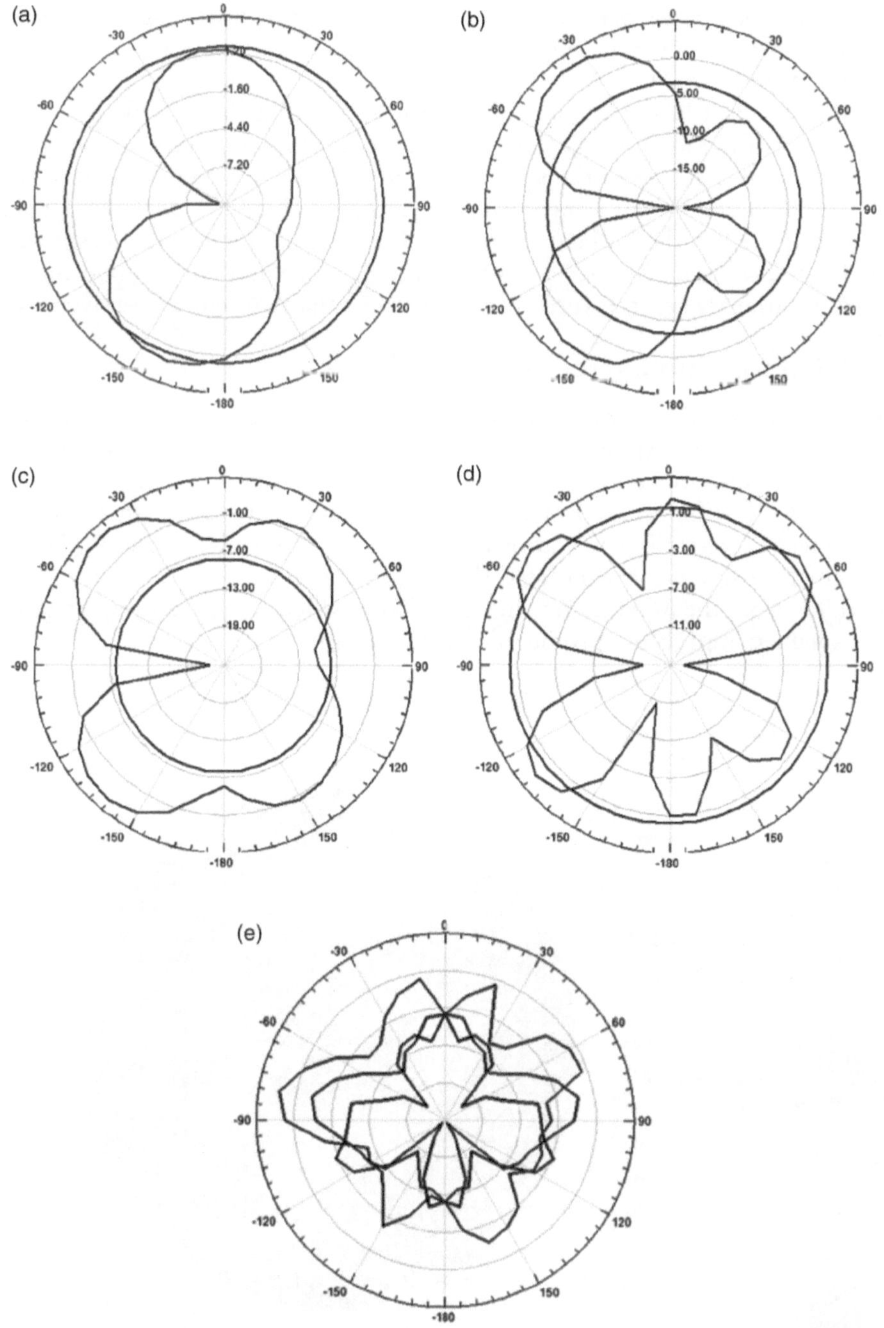

Figure 2.15 Proposed antenna 2-D radiation patterns measured at the following GHz frequencies: (a) $F_c = 2.4$; (b) $F_c = 5.3$; (c) $F_c = 6.7$; (d) $F_c = 8.2$; and (e) $F_c = 9.4$.

Input impedance is a vital parameter of an antenna. It describes the relationship between current and voltage at the antenna input [34]. The power radiated by the antenna or absorbed by the antenna circuit is represented by the real part of impedance, while power stored in the near field of the antenna is described by the imaginary part. It is important to remember that the power stored in the near field is nonradiated. A resonant antenna is one having real input impedance. Satisfactory values of input impedance were acquired by the FR4 substrate for all resonating frequencies, as shown in Table 2.6. It can be shown that good impedance matching between the feeding circuitry and the antenna results in enhanced antenna effectiveness.

2.4.3 Radiation pattern—2-D and 3-D

Figure 2.15 shows the far-field two-dimensional radiation patterns generated by an electromagnetic simulator for all five resonating frequencies, F_c, for each frequency band. Acceptable radiation characteristics are obtained in the H-plane for all the radiation patterns. The utility of the antenna for handheld or versatile applications is proved by uniformity in the E-plane. Figure 2.16 shows the simulated three-dimensional total gain for the second iteration; the maximum value is 3.27 dB, which is a good value of gain for such a compact low-profile microstrip antenna.

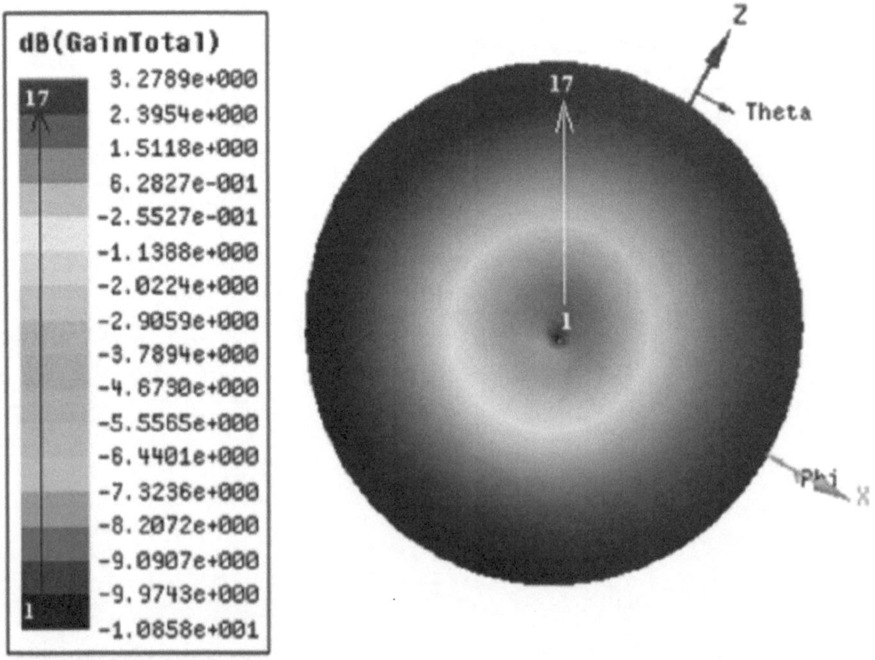

Figure 2.16 3-D total gain of the proposed antenna.

The proposed antenna demonstrates an omnidirectional radiation design with a figure eight shape similar to the radiation pattern of a dipole antenna. The actual antenna can be etched on a readily available and economical FR4 epoxy sheet.

2.5 CONCLUSION

Multiband fractal and hybrid antennas can be used widely for smart wireless networks and MIMO applications. Fractal shapes are complex, and it is not feasible to generate such shapes manually. But because these shapes can be defined by mathematical expressions, one can easily generate such complex shapes using IFS and scripting methods. Various fractal shapes can be modified by altering some parameters of their conventional counterparts. Such modified fractals provide enormous possibilities for researchers, and they can use shapes like radiating elements for planar antennas for multiband wireless applications. A pentaband planar microstrip hybrid fractal antenna has been designed on an FR4 substrate by integrating Koch curve and meander geometries. The hybrid fractal antenna geometry developed in this chapter was designed using the IFS method in MATLAB and importing the generated VBScript in an electromagnetic simulator. The proposed antenna exhibits multiband behavior at five different frequencies—2.441, 5.341, 6.743, 8.236, and 9.421 GHz—with acceptable values of input impedance and return loss. As it has a compact design, it is a perfect candidate for use in portable wireless communication devices.

REFERENCES

[1] C. Rowell and E. Y. Lam, "Mobile-phone antenna design," *IEEE Antennas and Propagation Magazine*, vol. 54, no. 4, pp. 14–34, 2012.

[2] S. Lee and Y. Sung, "Multiband antenna for wireless USB dongle applications," *IEEE Antennas and Wireless Propagation Letters*, vol. 10, pp. 25–28, 2011.

[3] L. Lizzi, R. Azaro, G. Oliveri, and A. Massa, "Printed UWB antenna operating over multiple mobile wireless standards," *IEEE Antennas and Wireless Propagation Letters*, vol. 10, pp. 1429–1432, 2011.

[4] S. R. Best, "On the resonant properties of the Koch fractal and other wire monopole antennas," *IEEE Antennas and Wireless Propagation Letters*, vol. 1, no. 1, pp. 74–76, 2002.

[5] S. R. Best, "On the performance properties of the Koch fractal and other bent wire monopoles," *IEEE Transactions on Antennas and Propagation*, vol. 51, no. 6, pp. 1292–1300, 2003.

[6] K. J. Vinoy, J. K. Abhraham, and V. K. Varadan, "On the relationship between fractal dimension and the performance of multi-resonant dipole antennas using Koch curves," *IEEE Transactions on Antenna and Propagation*, vol. 51, pp. 2296–2303, 2003.

[7] C. T. P. Song, P. S. Hall, and H. Ghafouri-Shiraz, "Perturbed Sierpinski multiband fractal antenna with improved feeding technique," *IEEE Transactions on Antennas and Propagation*, vol. 51, no. 5, pp. 1011–1017, 2003.
[8] C. Puente, J. Romeu, R. Pous, X. Garcia, and F. Benitez, "Fractal multiband antenna based on the Sierpinski gasket," *Electronics Letters*, vol. 32, no. 1, p. 1, 1996.
[9] C. P. Baliarda, C. B. Borau, M. N. Rodero, and J. R. Robert, "An iterative model for fractal antennas: Application to the Sierpinski gasket antenna," *IEEE Transactions on Antennas and Propagation*, vol. 48, no. 5, pp. 713–719, 2000.
[10] J. Robinson and Y. Rahmat-Samii, "Particle swarm optimization in electromagnetics," *IEEE Transactions on Antennas and Propagation*, vol. 52, no. 2, pp. 397–407, 2004.
[11] M. F. Pantoja, A. R. Bretones, F. G. Ruiz, S. G. Garcia, and R. G. Martin, "Particle-swarm optimization in antenna design: Optimization of log-periodic dipole arrays," *IEEE Antennas and Propagation Magazine*, vol. 49, no. 4, pp. 34–47, 2007.
[12] Anuradha, A. Patnaik, and S. N. Sinha, "Design of custom-made fractal multi-band antennas using ANN-PSO," *IEEE Antennas and Propagation Magazine*, vol. 53, no. 4, pp. 94–101, 2011.
[13] C. E. Balanis, *Antenna Theory: Analysis and Design*, 3rd ed. Hoboken, NJ: Wiley, 2005.
[14] R. Garg, P. Bhartia, I. Bahl, and A. Ittipiboon, *Microstrip Antenna Design Handbook*. Norwood, MA: Artech House, 2001.
[15] J. J. Carr, *Practical Antenna Handbook*, 4th ed. New York: McGraw-Hill, 2001.
[16] B. B. Mandelbrot, "The fractal geometry of nature," *American Journal of Physics*, vol. 51, no. 3, p. 286, 1983.
[17] Y. K. Choukiker and S. K. Behera, "Design of wideband fractal antenna with combination of fractal geometries," in *ICICS 2011—8th International Conference on Information, Communications and Signal Processing*, pp. 6–8, 2011.
[18] H. Oraizi and S. Hedayati, "Combined fractal geometries for the design of wide band microstrip antennas with circular polarization," in *IEEE 10th Mediterranean Microwave Symposium*, pp. 122–125, 2010.
[19] D. H. Werner and S. Ganguly, "An overview of fractal antenna engineering research," *IEEE Antennas and Propagation Magazine*, vol. 45, no. 1, pp. 38–57, 2003.
[20] R. Kyprianou, B. Yau, A. Alexopoulos, A. Verma, and B. D. Bates, "Investigations into Novel Multi-band Antenna Designs," pp. 1–27, 2006. https://citeseerx.ist.psu.edu/viewdoc/download?doi=10.1.1.591.5731&rep=rep1&type=pd
[21] C. B. Nsir, J. M. Ribero, C. Boussetta, and A. Gharsallah, "A wide band transparent Koch snowflake fractal antenna design for telecommunication applications," in *IEEE 19th Mediterranean Microwave Symposium*, pp. 1–3, 2019.
[22] P. Z. Petkov and B. G. Bonev, "Analysis of a modified Sierpinski gasket antenna for Wi-Fi applications," in *24th International Conference on Radioelektronika*, pp. 1–3, 2014.
[23] S. S. Khade and P. D. Bire, "Fractal MIMO antenna for wireless application," in *Optical and wireless technologies: Proceedings of OWT 2018*, ed. V. Janyani, G. Singh, M. Tiwari, and A. d'Alessandro, pp. 347–356. Singapore: Springer, 2020.

[24] H. A. M. Silva, A. G. D'Assuncao, and J. P. Silva, "Design of modified Pythagorean fractal antenna for multiband application," in *2017 International Applied Computational Electromagnetics Society Symposium*, pp. 1–2, 2017.
[25] Y. K. Choukiker, S. K. Sharma, and S. K. Behera, "Hybrid fractal shape planar monopole antenna covering multiband wireless communications with MIMO implementation for handheld mobile devices," *IEEE Transactions on Antennas and Propagation*, vol. 62, no. 3, pp. 1483–1488, 2014.
[26] Y. Kumar and S. Singh, "A compact multiband hybrid fractal antenna for multistandard mobile wireless applications," *Wireless Personal Communications*, vol. 84, no. 1, pp. 57–67, 2015.
[27] Y. Kumar and S. Singh, "Performance analysis of coaxial probe fed modified Sierpinski–Meander hybrid fractal heptaband antenna for future wireless communication networks," *Wireless Personal Communications*, vol. 94, no. 4, pp. 3251–3263, 2017.
[28] Y. Kumar and S. Singh, "A coaxial probe fed multiband hybrid fractal antenna for wireless applications," in *2016 International Conference on Wireless Communications, Signal Processing and Networking*, pp. 1717–1721, 2016.
[29] Y. Kumar and S. Singh, "A quad-band hybrid fractal antenna for wireless applications," in *2015 IEEE International Advance Computing Conference*, pp. 730–733, 2015.
[30] S. Tripathi, A. Mohan, and S. Yadav, "A compact Koch fractal UWB MIMO antenna with WLAN band-rejection," *IEEE Antennas and Wireless Propagation Letters*, vol. 14, pp. 1565–1568, 2015.
[31] S. Tripathi, S. Yadav, and A. Mohan, "Hexagonal fractal ultra-wideband antenna using Koch geometry with bandwidth enhancement," *IET Microwaves, Antennas and Propagation*, vol. 8, no. 15, pp. 1445–1450, 2014.
[32] H. Oraizi and S. Hedayati, "Microstrip multiband fractal dipole antennas using the combination of Sierpinski, Hilbert and Giuseppe Peano fractals," in *Proceedings of 2014 Mediterranean Microwave Symposium*, pp. 1–4, 2014.
[33] D. Punetha and V. Mehta, "An efficient progressive analysis on different dielectric substrates to design a rectangular microstrip patch antenna for mobile phones," in *2014 International Conference on Computational Intelligence and Communication Networks*, pp. 21–25, 2014.
[34] D. F. Mona and D. C. Nascimento, "Available universe of input impedances for the probe-fed circularly polarized microstrip antenna," in *2018 IEEE International Symposium on Antennas and Propagation & USNC/URSI National Radio Science Meeting*, pp. 107–108, 2018.

Chapter 3

Duplex antenna system for MIMO application

D. Venkata Siva Prasad, Harsh Verdhan Singh, Punya P. Pultani, and Shrivishal Tripathi

CONTENTS

DOI: 10.1201/9781003290230-3

3.1 INTRODUCTION

Wireless communication is a vital domain in science and technology. In wireless communication, the information is sent from one point to another in the form of electromagnetic radiation. The main device in wireless communication is the antenna, which acts like a transducer that is capable of converting the energy from one form to another—i.e., from the message signal in electrical form to electromagnetic radiation at the transmitter antenna and from electromagnetic radiation to electrical signal form at the receiver antenna. In modern wireless communication systems, mobile handsets are compact in size and light in weight, and this is made possible by employing efficient radiating elements. Patch antennas are compact and can be designed in different shapes and dimensions to accommodate mobile handsets. This advantage of the patch antenna makes it more suitable for mobile devices for wireless communication. The signal carrying information is transmitted from one device to another in wireless networks using standard protocols of wireless communication. To increase the throughput, the terminals have to communicate by simultaneously transmitting and receiving signals. The duplex antenna performs simultaneous transmission and reception of the signals between the terminals in wireless communication, and in general, a duplexer is defined as a device that supports two-way (duplex) communication over a common channel. The duplexer isolates the receiver and the transmitter ports of the system even though they share a common antenna elements for transmission and reception. In waveguides, the duplex functionality is performed by the devices like circulator and directional coupler. A duplexer's functionality can be divided based on frequency (filtering), polarization (orthogonality), or timing (switching) [1]. Duplex antennas have a significant role in the area of wireless communication as they double the spectral efficiency, which is essential for fourth and fifth generation (4G and 5G) standards of wireless communication. The mutual coupling is the main disadvantage in duplex antenna, which disturbs the gain and radiation pattern of the antenna. The applications of the duplex antenna in various domains and techniques to reduce mutual coupling are discussed in the chapter. Duplex communication systems are classified as half-duplex and full-duplex systems.

3.1.1 Half-duplex

A half-duplex system provides bidirectional communication and allows one to utilize the channel in only one direction at a time; one cannot utilize the channel in both directions simultaneously. Half-duplex systems need only one communication channel, which is used alternately for bidirectional communication by the shared antenna so that it conserves bandwidth. The walkie-talkie is a good example of a half-duplex system that uses only a single frequency channel for bidirectional communication [1].

3.1.2 Full-duplex

A full-duplex system, also called a double duplex, allows bidirectional communication between connected devices simultaneously in both directions. There is a technical difference between full-duplex communication and dual-simplex communication. In full-duplex communication, a single physical communication channel is used for both transmitting and receiving the signal. In dual-simplex communication, two different channels are used, one channel for transmitting and one channel for receiving [1]. The full-duplex operation is divided into two types of duplexing: (1) time-division duplexing and (2) frequency-division duplexing.

3.1.2.1 Time-division duplexing

In time-division duplexing, the transmitting or uplink signals and the receiving or downlink signals are transmitted over the same frequency band, but separate time slots are assigned for transmitting and receiving signals. Time-division duplexing provides greater trunk efficiency because the system utilizes the entire frequency band in the assigned time slot. A guard time slot is required to prevent interference between uplink and downlink signals. It imitates full-duplex communication over a single half-duplex communication channel.

3.1.2.2 Frequency-division duplexing

In frequency-division duplexing, the uplink and downlink signals are transmitted simultaneously using different frequency bands, which are separated by a guard band to avoid interference. Compared to time-division duplexing, frequency-division duplexing requires more bandwidth, but it has the advantage of not having to wait for the time slot to access the channel.

3.2 TYPES OF DUPLEXERS

There are different types of devices used in microwave engineering, radar antennas, wave guides, etc., to perform the duplexing operation. The functioning of different types of duplexers is explained here.

3.2.1 Transmit-receive switch

In radar, a transmit-receive (TR) switch performs the task of switching very rapidly between the transmitter and receiver, which share the same antenna, so that it provides the required isolation between receiver and

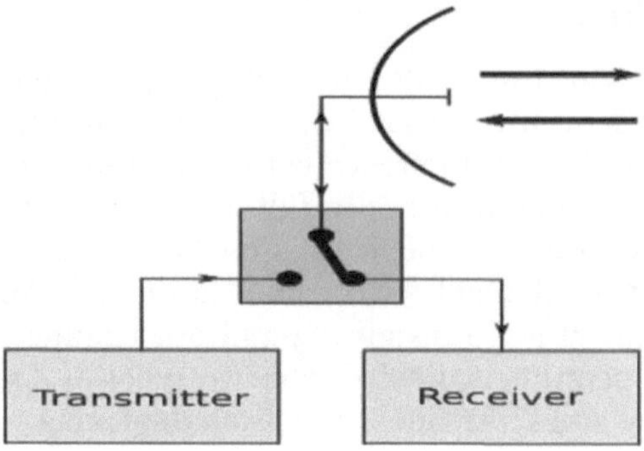

Figure 3.1 Transmit-receive duplexer.

transmitter. A TR duplexer, as shown in Figure 3.1, consists of a gas-discharge tube at the receiver input terminals. The tube conducts due to high voltage when the transmitter is activated, and the receiver terminals are shorted to the ground to protect it from the high voltage of the transmitter. The anti-transmit-receive (ATR) switch, which functions similarly to that of gas-discharge tube, disconnects the transmitter from the antenna when it is not transmitting any signal, so it prevents received energy from being wasted [1].

3.2.2 Circulator

A circulator is a multiport (three- or four-port), passive, nonreciprocal device as shown in Figure 3.2. The incident radio or microwave signal at any port couples to the next port in a circular direction [2]. In the case of a circulator with three ports, a signal incident at port 1 only comes out of port 2, and no signal appears at port 3. The applied signal at any particular port couples only to the adjacent port in rotation, and other ports are isolated. The three-port circulator is used to isolate the transmitter and receiver systems in duplex antenna.

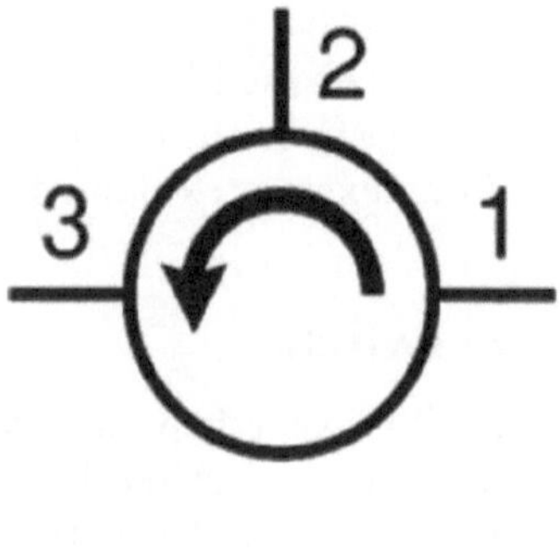

Figure 3.2 Circulator.

The scattering matrix [2] is

$$[S]=\begin{pmatrix} 0 & 0 & 1 \\ 1 & 0 & 0 \\ 0 & 1 & 0 \end{pmatrix} \tag{3.1}$$

Circulators are divided into two main categories depending on the materials used to fabricate them: ferrite circulators and nonferrite circulators.

3.2.2.1 Ferrite circulator

Ferrite circulators are composed of ferrite, which is a magnetic material, and they are radio-frequency circulators. They are divided into two main classes: one is a four-port waveguide circulator that works on the principle of Faraday rotation of electromagnetic waves that occur in a magnetized material, and the other is a three-port Y-junction circulator in which the propagating waves over two different paths get canceled near a magnetized material. In optical circulators, ferrimagnetic garnet crystal is used. Ferrite circulators have good "forward" signal circulation and greatly suppress the "reverse" signal circulation. The major disadvantage of ferrite circulators is their narrow bandwidth, and they will be bulky in size at low frequencies.

3.2.2.2 Nonferrite circulator

Nonferrite circulators use transistors that are nonreciprocal. Ferrite circulators are passive devices, whereas nonferrite circulators are active circulators that require external source of power to operate. The main drawbacks of an active circulator, that is transistor based, are the signal-to-noise degradation and the limitation of power. These characteristics are critical when the active circulator is used as a duplexer because it has to sustain the strong power of the transmit signal and the reception of the clean signal from the antenna.

3.2.3 Directional coupler

A directional coupler is a four-port waveguide, as shown in Figure 3.3, and it couples a small amount of power from the input signal for measurement purposes. In general, when port 1 is the input port, the applied signal is transmitted to port 2 (output port), and a small fraction of the power from the applied input signal at port 1, which is determined by the coupling factor of the directional coupler, is coupled to port 3 (coupled port). No signal appears (idle case) at port 4 which is known as the isolated or terminated port. However, in reality, a small portion of input signal power is obtained at the isolated port due to the mismatch of impedance at other ports, and

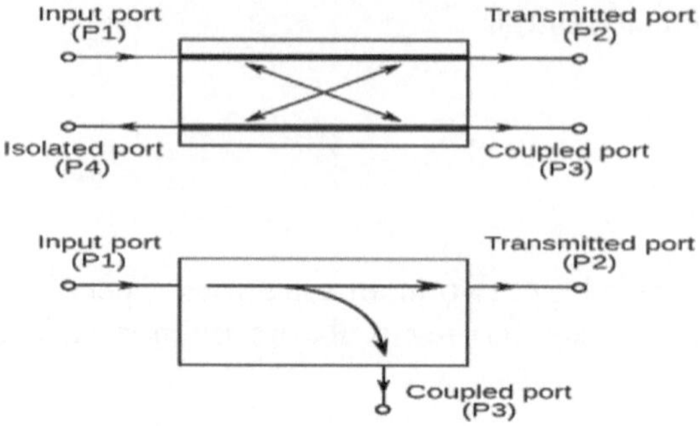

Figure 3.3 Directional coupler.

that reflected power at the isolated port is called back power that depends on the directivity of the coupler.

A hybrid coupler is a special case of the directional coupler in which the power of the input signal is split equally between the output port and coupled port. As half of the input signal power is obtained at the coupled port, the coupling factor of the device is 3 dB. There are two types of hybrid couplers. One is a quadrature hybrid coupler that gives a 90° phase shift between the signals at the output port and the coupled port when the signal is fed at the input port; it is an example of a symmetric coupler and has the following scattering matrix [2] form:

$$[S] = \frac{1}{\sqrt{2}} \begin{pmatrix} 0 & 1 & j & 0 \\ 1 & 0 & 0 & j \\ j & 0 & 0 & 1 \\ 0 & j & 1 & 0 \end{pmatrix} \tag{3.2}$$

The other type includes the rat-race and magic-T hybrid couplers that give a phase difference of 180° between the output signals at ports 2 and 3 when the input signal is fed at port 4; they are examples of antisymmetric couplers and have the following scattering matrix form [2]:

$$[S] = \frac{1}{\sqrt{2}} \begin{pmatrix} 0 & 1 & 1 & 0 \\ 1 & 0 & 0 & -1 \\ 1 & 0 & 0 & 1 \\ 0 & -1 & 1 & 0 \end{pmatrix} \tag{3.3}$$

3.3 MUTUAL COUPLING IN PATCH DUPLEXER ANTENNAS

Mutual coupling is mainly caused by three ways—by surface waves at the dielectric interface, surface currents on the ground plane, and near-field radiation between antennas. Mutual coupling produces induced surface currents on the other elements, causing impedance mismatching of the radiating elements and changing the radiation pattern of the antenna array. Some of the important techniques used to reduce mutual coupling are described next.

3.3.1 Electromagnetic bandgap structure

To decrease the strong mutual coupling between microstrip antennas that are on a thick substrate with a high value of permittivity, electromagnetic bandgap (EBG) structure is introduced between the radiating elements [3]. An EBG structure is a periodic structure that consists of metallic plates connected to the ground plane through vias, as shown in Figure 3.4. It acts like band-stop filter to suppress the surface waves in a particular frequency band. The compactness of an antenna [4, 5] is an important feature in communication antenna applications, and the mushroom-like EBG has better compactness when compared with other EBG structures that have dielectric rods and holes. Surface waves cause mutual coupling among the microstrip elements, and these surface waves at the resonating frequency

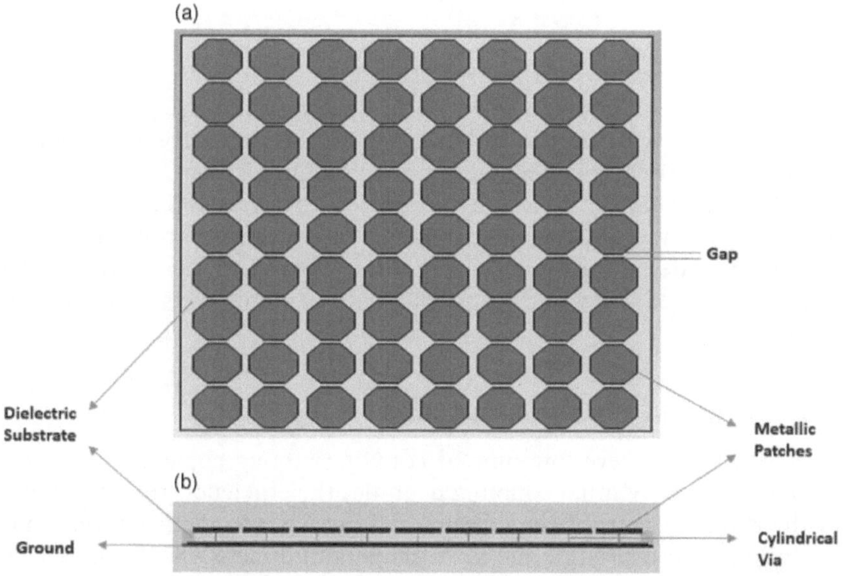

Figure 3.4 EBG structure: (a) top view; (b) side view.

of the antenna are well suppressed when the EBG structure parameters are designed properly, ensuring small mutual coupling between the radiating elements. The EBG structure improves the antenna's performance, as it increases the gain and decreases the back radiation of the antenna [6–8] when it suppresses the surface waves. Its structure comprises square patches arranged periodically, the substrate, and vias connecting the square patches to the ground plane. The structure has a stopband that mitigates the propagation of surface waves between the radiating elements at the resonant frequency of the radiating elements. An LC filter array is used as an example to explain the working of this structure: it acquires an inductive nature due to the current flowing from the metal patches to the ground plane through the vias and acquires a capacitive nature due to the gaps between adjacent metal patches. For the mushroom-like EBG structure, the parameters like the width of the patch, the gap between the patches, the thickness of the substrate, and the dielectric constant are denoted as W, g, h, and ε_r, respectively, and the capacitor (C) and inductor (L) values of the structure are determined using the following formulas [3]:

$$L = \mu_0 h \tag{3.4}$$

$$C = \frac{W\varepsilon_0 (1+\varepsilon_r)}{\pi} cosh^{-1}\left(\frac{2W+g}{g}\right) \tag{3.5}$$

$$f_0 = \frac{1}{2\pi\sqrt{LC}} \tag{3.6}$$

The resonant frequency of the EBG structure, given by Equation (3.6), is dependent on the L and C values, and at the resonant frequency f_0, the structure shows very high impedance, so the surface waves between the patch antennas are attenuated and mutual coupling is decreased. The resonating frequency of the structure depends on the dimensions of the patch and the gap between the patches and on the relative permittivity of the substrate. The design of the EBG structure is based on the condition that the resonating frequency of the antennas should match the stopband of the structure.

3.3.2 Defected ground structure

It is difficult to achieve low mutual coupling among coplanar microstrip antenna elements. Mutual coupling degrades the efficiency, the bandwidth, and the diversity gain of the antenna. Using a bandgap structure, which has filtering characteristics due to its shape and dimensions, is an efficient way to achieve low mutual coupling between antennas. The bandgap structure reduces the mutual coupling by attenuating the coupling fields and

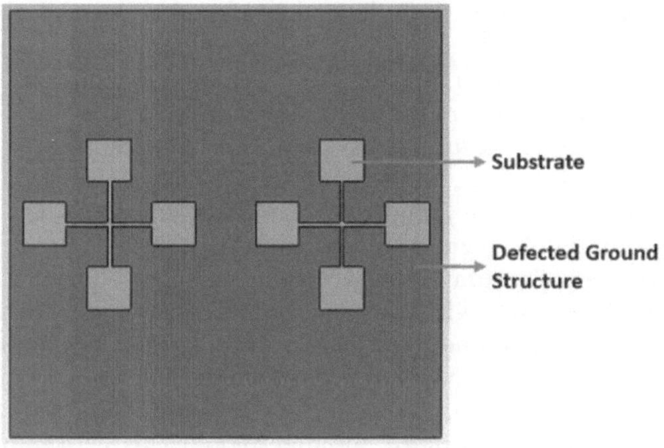

Figure 3.5 Defected ground structure.

inducing currents between the radiating elements. One of the most commonly used bandgap structures is the defected ground structure (DGS). A DGS is formed by etching a slot section on the ground plane, as shown in Figure 3.5. It obtains its capacitive and inductive nature due to the current flow along the slot edges. This makes the ground plane act like a bandgap structure. To achieve compactness of the antenna array, the dimensions of the planar bandgap structure filter should be minimum. Implementing fractals on the ground plane is an efficient way of minimizing the dimensions of the bandgap structure and also creating longer current lines on a smaller surface. The resonant frequency of the DGS depends on the length and width of the slot sections on the ground plane that determine both the effective capacitance and the effective inductance of the bandgap filter. The fractal DGS (FDGS) enhances the effective capacitance and inductance of the bandgap filter and also reduces the resonant frequency because it is inversely proportional to both capacitance and inductance [9]. The second and third iterative FDGSs are used to increase the isolation between the microstrip antenna elements. The third iterative FDGS shows better performance in terms of reducing mutual coupling than does the second iterative FDGS. FDGSs at higher-level iterative decrease the dimensions of the antenna [10]. The induced surface current on the idle antenna that is adjacent to the radiating one is reduced to the maximum extent by using an FDGS between them.

3.3.3 Decoupling network

In a mobile terminal, a multiple antenna system is used to enhance its performance, but placing multiple antennas in a compact area results in intense mutual coupling among the elements of the array and increases in pattern/spatial correlation, which degrades the channel capacity. A part of the power that feeds to one of the antennas will be wasted in the form of mutual coupling with the other antenna instead of radiating to free space, and this reduces the signal-to-noise ratio as well as the radiation efficiency of the antenna system. This degradation in the signal-to-noise ratio and radiation efficiency further deteriorates the channel capacity [11]. Mutual coupling also disturbs the impedance matching among the radiating elements and ports of the antenna system. A decoupling network is incorporated between the radiating elements to diminish the mutual coupling without disturbing the input impedance of the antenna. The decoupling network is a structure made by combining elements or striplines with a particular design and dimensions, as shown in Figure 3.6. A coupled resonator decoupling network (CRDN) is one of the decoupling networks used in an antenna array [12–13]. It enhances the efficiency of the individual elements in the array by matching their respective input impedance with the feed line using open-circuited stubs of the proper dimensions at the proper positions. A CRDN not only matches the impedance of the individual antennas but also suppresses the coefficients of coupling between the antennas with the help of a particular structure of decoupling striplines. It reduces the correlation and increases the radiation efficiency significantly, which improves the MIMO antenna's channel capacity as well as its throughput. It can be used in both asymmetric and symmetric arrays of antennas.

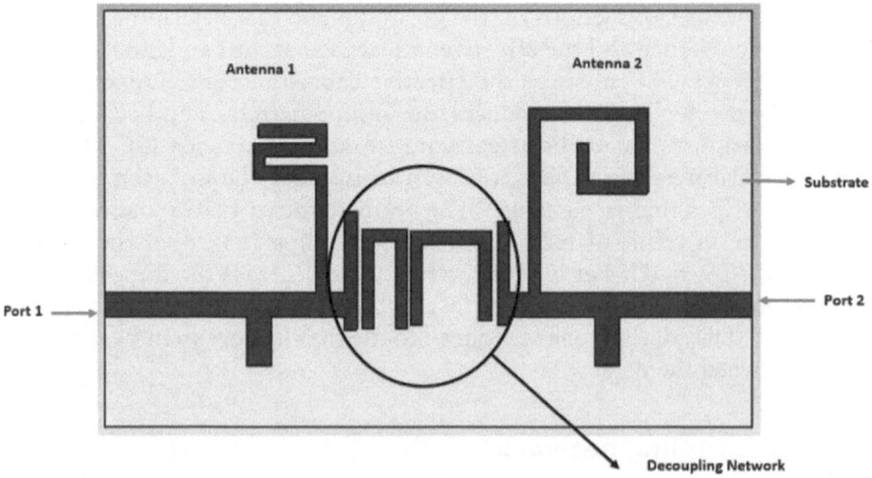

Figure 3.6 Decoupling network in an asymmetric array.

3.4 APPLICATIONS OF DUPLEXER ANTENNAS

Duplexer antenna have many applications, especially in 5G wireless communications, the internet of things (IoT), cellular networks, and mobile handsets. Full-duplex radios enable simultaneous transmission and reception of the information signal and also improve spectral efficiency by using the same channel. Full-duplex has the potential to double the efficiency of the spectrum because of concurrent signal transmission and reception when compared to conventional half-duplex mode.

3.4.1 5G communications in frequency range 1 (< 6 GHz)

A new generation of networks are used in 5G communications that improve the performance of wireless communication system in terms of energy efficiency, throughput, coverage, and spectral efficiency [14]. The 5G communication system, in general, operates in two frequency bands—frequency range 1 and frequency range 2. In frequency range 1, the operating frequencies are below 6 GHz, and in frequency range 2, the operating frequencies are above 24 GHz. In an application, the frequency band of 3–4 GHz is dedicated to full-duplex operation in the MIMO antenna system for 5G communications [15]. The full-duplex antenna intended to operate at dual-band consists of two radiating patches of certain dimensions suitable to resonate at the center frequency of their respective band. Each rectangular patch is connected to the corresponding port through the microstrip fedline to achieve dual-band full-duplex operation, as shown in Figure 3.7. The single-input single-output (SISO) system consists of two square patch antennas used together for transmitting and one square patch antenna for receiving; these transmitter and receiver antennas are placed close to each other [16]. In the transmitter, the two patches are fed with a phase shift of 180° and equal power using a power divider network in order to have maximum radiation in the broadside direction of the antenna system. Even though the transmitter and receiver patches are near each other, good isolation has been obtained because the surface currents on the transmitter and receiver patches are orthogonal to each other. This arrangement of patch antennas achieves good polarization of orthogonal diversity. Two such SISO subsystems are positioned orthogonally to realize dual polarization and improve the isolation between ports of the MIMO antenna system.

A MIMO antenna performing in half-duplex mode for 4G/LTE (Long-Term Evolution) and full-duplex mode for 5G communications is proposed in [17]. The multimode mobile terminals that operate in two bands are needed to support both 4G/LTE and 5G communications simultaneously because the transition from 4G to 5G is not rapid. There are two frequency bands for 5G wireless communications: one band is below 6 GHz, and

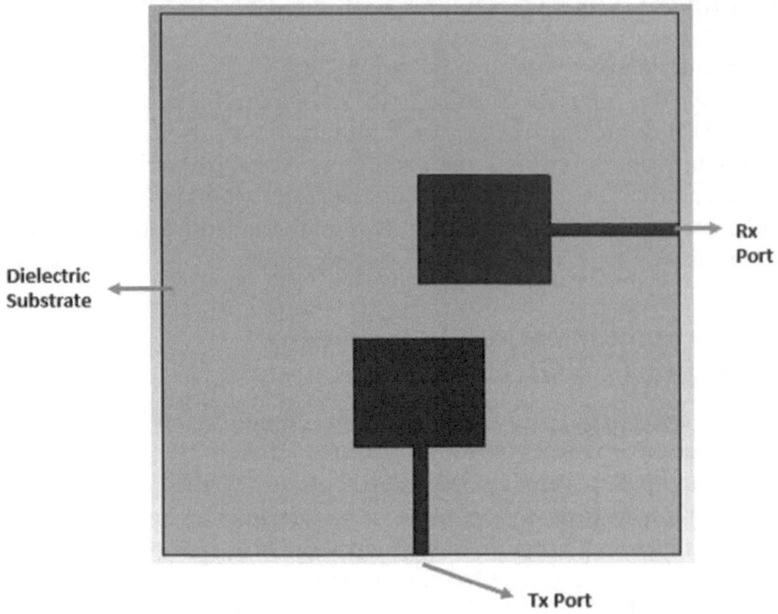

Figure 3.7 Full-duplex antenna system.

the other band is above 24 GHz. The planar inverted-F antenna (PIFA) structure consists of a rectangular patch antenna over a ground plane supported by two metal structures, one for feeding the antenna and the other for grounding. The PIFA is designed with the proper dimensions so that it resonates at the frequency band allotted to 4G/LTE communications. To obtain dual-band functionality, the surface of the PIFA is slotted in such a way that it resonates at the frequency band allotted to 5G communications. The slot section on the PIFA is etched at a minimum current distribution area to maintain good resonance at the lower band. The slot design is chosen so that the PIFA will have minimum return loss and maximum radiation efficiency. A parasitic element is also added to widen the higher band. An efficient two-PIFA system that works at a 4G band of 2.5–2.7 GHz and a 5G band of 3.4–3.8 GHz is presented in [18]. The two PIFA antennas are arranged at the corner edges of the ground plane in three distinct configurations, determined by the symmetry of the minor axis, the major axis, and the center of the ground plane. Among these three configurations, symmetry based on the center of the ground plane gives maximum separation between the antennas. The position of the two antennas on the ground plane is chosen so that there is a high degree of isolation between them.

In MIMO antennas, high efficiency, improved isolation, and enhanced gain are obtained by using dielectric resonator antennas (DRAs), as shown

Figure 3.8 Dielectric resonator antenna.

in Figure 3.8. A high degree of isolation between antenna elements is an important requirement in the MIMO antenna system. Much research has been done using such techniques as hybrid feeding, metasurface shields [19], and frequency selective surfaces (FSSs) [20] to get a sufficient degree of isolation between MIMO DRAs. The current displacement is resisted or attenuated by these techniques within antenna elements. The metal strips are added to the upper surface of the dielectric resonator to further increase the isolation between ports. The metal strips placed on the upper surface of the DRA are coupled electromagnetically with the fed lines through the slot section inserted in the ground plane between the feeding network and the dielectric. This arrangement improves diversity gain as well as channel capacity.

Reconfigurable antennas that are compact in size and simple in structure and that have better polarization performance and stability in radiation pattern are preferred over traditional antennas. The significant way to minimize antenna size is to employ a multiband MIMO antenna. Recently, the metamaterial-based system has been used to increase the isolation among antennas in recent times. One such decoupling method using a metasurface is proposed in [21] to improve isolation and radiation efficiency. A low-profile metasurface-based antenna is required for a wideband application that works in the S-band. By incorporating a partially reflective surface (PRS) and metasurface, the structure of the shared aperture antenna is used for dual-band operation, particularly at the S-band and Ka-band [22]. The dual-band aperture antenna gives optimum gain and maximum isolation. The changing nature of the physical and electrical properties of an antenna is the benefit of a shared surface that permits resonant modes at the S-band and Ka-band [23].

The mutual coupling in closely arranged antenna pairs will be very strong, and a method called the orthogonal-mode method is followed to mitigate the mutual coupling in such a scenario without using any external decoupling structure. The MIMO system proposed in [24] provides a promising solution to overcome the adverse effects due to mutual coupling in the compact MIMO antennas of 5G mobile phone and also to provide good isolation between antennas and better diversity performance. Several methods are proposed to enhance the isolation among closely spaced radiators. One method involves inserting a new coupling path between two antennas to wipe out the principal coupling currents, and typical methods use the neutralization line or decoupling element [25]. The orthogonal-mode method is a natural decoupling technique that does not require any additional structure to suppress the coupling currents between antennas [26]. In the dipole antenna, one edge is connected to the fed line, and the other edge is connected to the ground through shorting vias. When either of the two antennas is excited, keeping the other matched, it is observed that the surface current distributions are orthogonal on the antennas as well as on the ground plane. The following two sets of orthogonal modes perform the decoupling between tightly arranged antenna pairs: (1) the spatial coupling can be reduced by the distribution of currents orthogonally to the antenna pair, and (2) the coupling through the ground current can be reduced by the orthogonal distribution of currents over the ground plane.

The folded monopole antenna elements are present in the MIMO antenna, and they are placed symmetrically and orthogonally at the corners of the frame of the substrate; they are located close to each other without clearance on the ground plane [27]. So the antenna will resonate at dual bands, the arm of the monopole is folded. By properly designing and optimizing the dimensions of the folded monopole antenna's arm, the antenna can resonate at the desired bands of 3.4–3.6 GHz and 4.55–4.75 GHz, which are used in applications of 5G mobile handsets. To reduce the mutual coupling, a neutral line is inserted; it can be very short because the structure of the radiating element is optimized and the antenna can be compact. The isolation and the efficiency in dual bands are increased by the optimized structure of the antenna and the neutral line.

A dual-polarized, dual-differential 2.4 GHz microstrip square patch antenna with four ports is discussed in [28]. It is excited differentially by two identical3 dB/180° hybrid couplers for both transmitting and receiving operations. The square patch is used as the radiating antenna with four ports on its four sides, and each port of the patch is excited by a thin $\lambda/4$ microstrip feed line at the center of its respective edge. Both transmitting and receiving operation modes are excited by the differential fed mechanism through a pair of ports that are placed opposite to each other;

that is ports 1 and 3 for the transmitting operation and ports 2 and 4 for the receiving operation in differential mode. While in transmitting mode, ports 1 and 3 are excited with equal power and a phase shift of 180° by the hybrid coupler. Ports 2 and 4 are connected to the receiver by a coupler that receives approximately equal amounts of leakage power from the transmitter ports, which cancel out due to the differential mechanism of the coupler at the receiver port. Because of this, a good amount of isolation is obtained between the transmitter and receiver terminals. The patch resonates at dual-band frequencies, one band for transmitting and the other band for receiving, with linear orthogonal polarization that is achieved by placing the transmitter and receiver ports perpendicular to each other [29].

In radio communications, the efficiency of bandwidth utilization can be improved by employing single-frequency full-duplex (SFD) wireless communication [30]. The essential requirement to realize SFD is a high degree of isolation between the receiver and transmitter so that a radio signal can be received from the intended transmitter without substantial deterioration of the signal-to-interference ratio produced by self-interference from the transmitter [31]. Such inherent isolation can be obtained in collinear dipole arrays by placing each antenna in the null direction of the other's radiation pattern. However, the isolation achieved by this natural phenomenon is not sufficient in full-duplex operations and requires greater isolation between the antennas to increase the signal-to-interference ratio required for the smooth functioning of a full-duplex antenna. The isolation is increased by using a parasitic coplanar decoupling structure in the collinear dipole array [32]. To improve the isolation, the radiating elements of the decoupling structure should have polarization that is orthogonal to that of the collinear transmitter and receiver dipoles. The analysis shows that the decoupling structure produces fields that are out of phase or counter to the original ones around the receiver dipole.

3.4.2 5G communications in frequency range 2 (> 24 GHz)

In recent years, millimeter-wave (mmWave) wireless communication has become a significant area in wireless communications beyond 5G. Compared to the existing LTE 2.5 GHz band, the atmospheric attenuation and free space path loss (FSPL) are considerably higher in mmWave wireless communication. For acceptable transmission distances, a high-gain antenna is required for the mmWave wireless communication system. Reflector antennas that are capable of giving high gains have been used in several mmWave and terahertz wireless communication systems for the past few years, and these high-gain reflector antennas can attain data rates up to tens of Gbps because the data rates are proportional to the gain of the antenna [33, 34]. The antenna system consists of a feed and a dual-reflector system based on an axially displaced ellipse (ADE) structure [35]. The feed is capable of generating both right-hand circular polarization (RHCP) and left-hand circular polarization (LHCP) signals by using a stepped septum polarizer. The high gain in radiation is obtained by illuminating both the sub-reflector and the main reflector by stepped septum polarizer with ring focuses. An ADE reflector antenna is also designed to give a minimum back reflection to the feed maintaining the high gain of the antenna. Polarization-division multiplexing (PDM) is employed by using two circular polarized signals that are orthogonal to each other to simultaneously receive and transmit the signal to achieve full-duplex operation in mmWave wireless communication systems. High isolation and low axial ratio (AR) are achieved by properly designing and optimizing the antenna feed having a bandwidth range from 75 to 95 GHz.

3.4.3 Wireless local area network systems

The polarization of radio wave changes very rapidly and significantly in urban and indoor communication environments due to multiple reflections and diffractions by objects between the transmitter and receiver antennas [36]. A horizontally polarized antenna is one of the solutions to mitigate the polarization variations in environments that are rich in multipath propagation. The antenna with an omnidirectional pattern and wide spatial coverage is appropriate in wireless local area network systems (WLANs) [37, 38]. The antenna system in which both transmitter and receiver antennas have horizontal polarization is known as a co-horizontally polarized antenna and is called an in-band full-duplex (IBFD) antenna system if the same frequency is used for both transmitting and receiving in the full-duplex operation. In WLANs, co-horizontally polarized IBFD antennas with omnidirectional patterns are used. A variant turnstile antenna with four T-shaped monopoles that are 90° progressively phased is used as a transmitter antenna, whereas an in-phase loop antenna that consists of four dipoles that are symmetrically excited is used as a receiver antenna [39]. Both transmitter and receiver antennas give omnidirectional radiation patterns in the azimuth plane and are horizontally polarized. Due to an orthogonal spatial phase, the mutual coupling among the antennas is reduced to a maximum extent although the two antennas are co-polarized and close to each other, and, theoretically, the isolation between ports is infinite.

A monopole antenna that is used in a dual-band full-duplex system for a WLAN application is proposed in [40]. To achieve higher isolation in the preferred frequency bands, the receiver ports of the transceiver antenna are fed through a 180° ring hybrid coupler, which is a dual-band coupler. For both transmitting and receiving antennas to function in dual bands, extra arms are added to the respective single monopoles, as shown in Figure 3.9. In receiving antenna, an L-shaped arm is inserted on one side of the monopole, whereas in the transmitting antenna, symmetry is

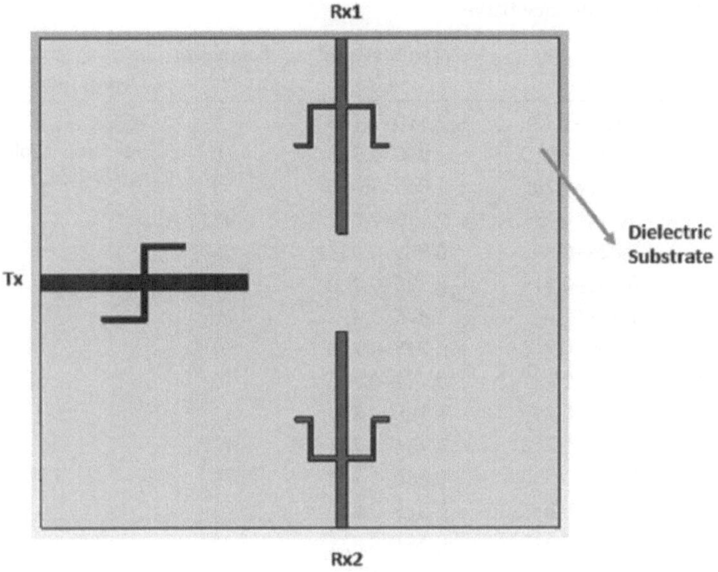

Figure 3.9 Dual-band monopole antenna.

maintained by adding L-shaped arms on two sides of the monopole [41]. Because of the symmetry of the transmitter antenna, both receiver antennas get equal leakage power. The coupled transmitter leakage power will get canceled at the receiver port because of the differential ports of the coupler. Due to the orientation of the receiving antenna and the phase shift of 180° introduced by the coupler, the received power from the two receiver antennas will get added. In this way, isolation is improved between ports by the coupler and the orientation of the transmitter and receiver antennas.

3.4.4 Internet of things networks

The IoT is a network that connects home appliances, vehicles, physical devices, and other items with electronic systems, actuators, sensors, software, and network connectivity, and this allows the things (physical devices) to connect and exchange data. The embedded computing system gives each thing in the network a unique identity and enables the devices to interact with the help of the existing Internet infrastructure. Narrow band-IoT (NB-IoT) is a narrowband radio technology intended for the IoT and standardized by the 3rd Generation Partnership Project (3GPP). The list of bands and their mode of duplex operation used in NB-IoT is shown in the Table 3.1. This standard enables a large number of connected devices and focuses on indoor coverage and long battery life [42].

Table 3.1 NB-IoT frequency bands

NB-IoT band	*Uplink band (GHz)*	*Downlink band (GHz)*	*Bandwidth (MHz)*	*Duplex mode*
B1	1.920–1.980	2.110–2.170	60	Half-duplex frequency division duplexing (HD-FDD)
B2	1.850–1.910	1.930–1.990	60	
B3	1.710–1.785	1.805–1.880	75	
B4	1.710–.755	2.110–2.155	45	
B5	0.824–0.849	0.869–0.894	25	
B8	0.880–0.915	0.925–0.960	25	
B11	1.427–1.447	1.475–1.495	20	
B12	0.699–0.716	0.729–0.746	17	
B13	0.777–0.787	0.746–0.756	10	
B14	0.788–0.798	0.758–0.768	10	
B17	0.704–0.716	0.734–0.746	12	
B18	0.815–0.830	0.860–0.875	15	
B19	0.830–0.845	0.875–0.890	15	
B20	0.832–0.862	0.791–0.821	30	
B25	1.850–1.915	1.930–1.995	65	
B26	0.814–0.849	0.859–0.894	35	
B28	0.703–0.748	0.758–0.803	45	
B31	0.452–0.457	0.462–0.467	5	
B66	1.710–1.780	2.110–2.200	70/90	
B70	1.695–1.710	1.995–2.020	25	
B71	0.633–0.698	0.617–0.783	65	
B72	0.451–0.456	0.461–0.466	5	
B73	0.450–0.455	0.461–0.466	5	
B74	1.427–1.470	1.475–1.518	43	
B85	0.698–0.716	0.728–0.746	10	

3.4.5 Cellular networks

In IBFD wireless communication systems, the radios transmit and receive the signals on the same frequency band simultaneously, and good duplexing performance is due to recent improvements in self-interference cancelation techniques. The spectrum efficiency is doubled by the duplexing operation of the IBFD system at the physical layer of the wireless network. Such a system also solves some important problems such as large end-to-end delays, hidden terminals, and loss of throughput due to congestion in existing wireless networks. Self-interference cancellation techniques can be implemented in the following three domains: (1) The propagation domain deals at the antenna level, where self-interference is canceled by implementing proper antenna design and by using polarization and pattern-diversity methods. (2) The analog-circuit domain is where self-interference is canceled using analog circuits such as attenuators and phase shifters between two ports.

(3) The digital domain is where self-interference is canceled using digital signal processing methods [43]. Self-interference cancellation technologies are broadly divided into two categories: one is passive suppression, where the transmitter and receiver antennas are isolated electromagnetically, and the other is active cancellation, in which the self-interference is mitigated by using a node's knowledge of its own transmitted signal [44]. Generally, propagation-domain suppression can be achieved in different ways—e.g., directional antennas, transmit-beamforming, antenna orientation, path loss, and duplexers.

A full-duplex cellular network consists of a full-duplex base station (FDBS) that supports a set of half-duplex down-link (HDDL) users and a set of half-duplex up-link (HDUL) users [45]. The transmitter and the receiver of the base station consist of reconfigurable antennas, and each reconfigurable antenna selects the mode for transmitting or receiving from several preset modes. One of the possible scenarios for full-duplex radios that are compatible with half-duplex users in the current communication systems is that where the FDBS concurrently supports a section of HDDL users and a section of HDUL users. Interference from HDUL users to HDDL users is mitigated by orthogonalizing the HDDL and HDUL traffic in the frequency or time domain. This interference does not occur in half-duplex cellular networks. The network throughput is degraded if interference from HDUL to HDDL users is not well attenuated, even though self-interference is suppressed perfectly. The key challenge in boosting the network throughput is to mitigate interference from HDUL to HDDL users efficiently by adjusting full-duplex operation at the base station.

Having the signal-to-noise ratio at its maximum value at the input of the receiver is a condition for good performance of a full-duplex radio device, which is called an IBFD radio if uses a single channel for transmitting and receiving. A large signal-to-noise ratio at the receiver input is possible when the transmitter signal power has minimum impact. Between 90 and 120 dB of self-interference must be canceled for the smooth working of a full-duplex device. This is achieved by using both digital and circuit techniques to suppress the self-interference. In the circuit techniques, at least 60 dB of self-interference must be suppressed to avoid the saturation of the receiver circuit [46]. To obtain the high degree of isolation required for full-duplex communication, a dual-polarized antenna is used in which the polarizations of the transmitter and receiver antenna are orthogonal. The isolation of ports can be further improved by using a feding technique that is anti-phased by pairwise repetition through successive mirror-image and 180° phase-shifted driving currents of the whole radiator group (two, four, etc.) in the array of multi-element antenna that uses a corporate feeding network (CFN) having many equal power distributions [47]. The radiating element that is coupled to the feed line electromagnetically through the slot section when used as a single antenna in the array of multi-element and multilayer antennas gives a wide operating bandwidth.

3.4.6 Wireless communication and frequency modulated continuous wave radar systems

The subarray cavity-backed slot antenna (CBSA), which is a common-aperture antenna with broad bandwidth and dual polarization, has a very high degree of isolation between two ports, and this is a suitable condition for full-duplex operation [48]. The antenna is intended for wideband full-duplex operation in wireless communication and frequency modulated continuous wave (FMCW) radar systems. The antenna system consists of a thin rectangular cavity and four cross-slots that are cut out on the cavity for radiation purposes [49]. Two concentric orthogonal slots that are on the back wall of the rectangular cavity are used to feed the cross-slots for radiating. An end-launch coaxial-to-waveguide transition, which is used to excite the cavity through a feeding slot, is used to excite one polarization. The other polarization is excited by a two-pronged microstrip line that crosses over another, symmetrical feeding slot and is oriented orthogonally to the feeding slot. The minimum mutual coupling is obtained over a wide bandwidth between channels by designing a microstrip line in such a way that each prong has an out-of-phase coupling with the waveguide feed.

3.4.7 Mobile applications

There are three domains where signal interference is suppressed. The first is the propagation domain, where most of the signal interference suppression is done at the antenna stage; this reduced the amount of signal interference suppression that must be done in the analog circuit and digital domains. The single-patch antenna is used as a duplexer by implementing the dual-polarization technique, and the duplexing performance is improved by using a feed-forward loop that consists of a 10 dB coupler, attenuator, and phase shifter between the transmitting and receiving ports, as shown in Figure 3.10. The interference signal from the transmitter is cancelled by the out of phase signal from the feed forward loop at the receiver port [50]. This arrangement gives a maximum interport isolation at the center frequency. Even minor variations in loop attenuation and phase shift cause a huge change in the isolation between ports.

For small-form-factor devices, designing multiple antennas with a high degree of isolation is very challenging due to the strict limitation on space and size. Currents on the surface of the ground plane are the main cause of coupling between the multiple antenna elements. Excitation of the currents orthogonally on the ground plane that leads to pattern diversity can be utilized to achieve isolation of the antennas. A combination of balanced antennas with capacitive coupling elements is used to achieve pattern diversity. The theory of characteristic modes (TCM) is used to find an optimum location for the coupling elements to excite a single mode [51]. Self-resonant

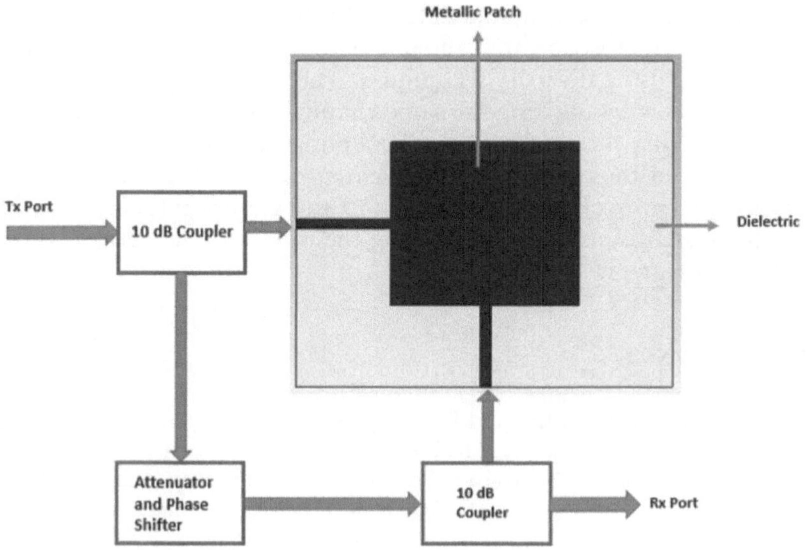

Figure 3.10 Duplex Antenna with feed-forward loop.

balanced antennas are used to localize the flow of current on the ground plane. A three-port antenna system consisting of two balanced dipoles (TX antennas) and coupling elements (RX antennas) is presented in [52]. Two balanced dipoles are located on the short edges of the chassis. To excite the dipole antennas, two T-slots are added in the ground plane. Two capacitive coupling elements are located at the center of the long edge of the chassis, since the full-wavelength current of the chassis has its minimum values at the center of the chassis. A simple T-splitter is used as a feeding network to excite the coupling elements in-phase. In this manner, selective excitation of the full-wavelength current mode of the chassis can be achieved. On the other hand, the balanced dipole antennas excite the half-wavelength current flowing on the short edge of the chassis. The dipole elements are assigned as TX antennas, and port 3 is assigned to the RX antenna. The S-parameters of the antenna are simulated in different scenarios like free space (FS); talking mode, which is referred to as head to hand (HH); data mode, which is referred to as one hand (OH); and landscape mode, which is referred to as two hands (TH).

3.5 SUMMARY

In this chapter, the workings of duplex devices and different types of duplexers, with relevant scattering matrix forms and types of duplexing operations, are discussed. The two-way communication between two devices is made easy by employing a duplex antenna where the duplexer device is used to share a common antenna for both transmitter and receiver terminals. The microstrip patch antenna makes the duplexer antenna compact in size by using diversity techniques. The duplex operation increases spectral efficiency by utilizing the same channel concurrently for transmitting and receiving the signal. Mutual coupling decreases the efficiency of the antenna and thus reduces the duplexing operation. Different designs and techniques to increase the isolation of the duplex antenna in various applications are discussed. By increasing the isolation, the performance of the duplex antenna is improved.

REFERENCES

[1] Kenneth E Kolodziej. In-Band Full-Duplex Wireless Systems Handbook, 3rd ed. ISBN:9781630817893, Artech House.
[2] Pozar, D. M. (2011). Microwave Engineering, 4th ed. Hoboken, NJ: Wiley.
[3] Yang, F., & Rahmat-Samii, Y. (2003). Microstrip antennas integrated with electromagnetic band-gap (EBG) structures: A low mutual coupling design for array applications. IEEE Transactions on Antennas and Propagation, 51(10), 2936–2946.
[4] Yang, F., Kee, C. S., & Rahmat-Samii, Y. (2001, July). Step-like structure and EBG structure to improve the performance of patch antennas on high dielectric substrate. In IEEE Antennas and Propagation Society International Symposium. 2001 Digest. Held in conjunction with: USNC/URSI National Radio Science Meeting (Vol. 2, pp. 482–485). IEEE.
[5] Gonzalo, R., De Maagt, P., & Sorolla, M. (1999). Enhanced patch-antenna performance by suppressing surface waves using photonic-bandgap substrates. IEEE Transactions on Microwave Theory and Techniques, 47(11), 2131–2138.
[6] Yang, F., & Rahmat-Samii, Y. (2001, July). Mutual coupling reduction of microstrip antennas using electromagnetic band-gap structure. In IEEE Antennas and Propagation Society International Symposium. 2001 Digest. Held in conjunction with: USNC/URSI National Radio Science Meeting (Vol. 2, pp. 478–481). IEEE.
[7] Rahman, M., & Stuchly, M. A. (2001, July). Wideband microstrip patch antenna with planar PBG structure. In IEEE Antennas and Propagation Society International Symposium. 2001 Digest. Held in conjunction with: USNC/URSI National Radio Science Meeting (Vol. 2, pp. 486–489). IEEE.
[8] Sharma, S. K., & Shafai, L. (2001, July). Enhanced performance of an aperture-coupled rectangular microstrip antenna on a simplified unipolar compact photonic band gap (UC-PBG) structure. In IEEE Antennas and Propagation Society International Symposium. 2001 Digest. Held in conjunction with: USNC/URSI National Radio Science Meeting (Vol. 2, pp. 498–501). IEEE.
[9] Hammoodi, A. I., Al-Rizzo, H. M., & Isaac, A. A. (2015, July). Mutual coupling reduction between two monopole antennas using fractal based DGS. In 2015 IEEE International Symposium on Antennas and Propagation & USNC/URSI National Radio Science Meeting (pp. 416–417). IEEE.

[10] Wei, K., Li, J. Y., Wang, L., Xing, Z. J., & Xu, R. (2016). Mutual coupling reduction by novel fractal defected ground structure bandgap filter. IEEE Transactions on Antennas and Propagation, 64(10), 4328–4335.
[11] Yun, J. X., & Vaughan, R. G. (2011). Multiple element antenna efficiency and its impact on diversity and capacity. IEEE Transactions on Antennas and Propagation, 60(2), 529–539.
[12] Zhao, L., Yeung, L. K., & Wu, K. L. (2012, December). A novel second-order decoupling network for two-element compact antenna arrays. In 2012 Asia Pacific Microwave Conference Proceedings (pp. 1172–1174). IEEE.
[13] Zhao, L., Yeung, L. K., & Wu, K. L. (2014). A coupled resonator decoupling network for two-element compact antenna arrays in mobile terminals. IEEE Transactions on Antennas and Propagation, 62(5), 2767–2776.
[14] 5G White Paper, 5G Vision, Requirements, and Enabling Technologies, March 2019. https://trai.gov.in/sites/default/files/White_Paper_22022019.pdf
[15] Gbafa, K., Diallol, A., Le Thuc, P., & Staraj, R. (2018, July). Tx/Rx antenna system for full-duplex application. In 2018 IEEE International Symposium on Antennas and Propagation & USNC/URSI National Radio Science Meeting (pp. 1571–1572). IEEE.
[16] Gbafa, K., Diallo, A., Le Thuc, P., & Staraj, R. (2019, October). High isolated MIMO antenna system for full-duplex 5G applications. In 2019 IEEE Conference on Antenna Measurements and Applications (pp. 49–52). IEEE.
[17] Bhatti, R. A., Im, Y. T., & Park, S. O. (2009). Compact PIFA for mobile terminals supporting multiple cellular and non-cellular standards. IEEE Transactions on Antennas and Propagation, 57(9), 2534–2540.
[18] Fakih, M. A., Diallo, A., Le Thuc, P., Staraj, R., Mourad, O., & Rachid, E. A. (2019). Optimization of efficient dual band PIFA system for MIMO half-duplex 4G/LTE and full-duplex 5G communications. IEEE Access, 7, 128881–128895.
[19] Dadgarpour, A., Zarghooni, B., Virdee, B. S., Denidni, T. A., & Kishk, A. A. (2016). Mutual coupling reduction in dielectric resonator antennas using metasurface shield for 60-GHz MIMO systems. IEEE Antennas and Wireless Propagation Letters, 16, 477–480.
[20] Karimian, R., Kesavan, A., Nedil, M., & Denidni, T. A. (2016). Low-mutual-coupling 60-GHz MIMO antenna system with frequency selective surface wall. IEEE Antennas and Wireless Propagation Letters, 16, 373–376.
[21] Liu, F., Guo, J., Zhao, L., Huang, G. L., Li, Y., & Yin, Y. (2019). Dual-band metasurface-based decoupling method for two closely packed dual-band antennas. IEEE Transactions on Antennas and Propagation, 68(1), 552–557.
[22] Li, T., & Chen, Z. N. (2019). Shared-surface dual-band antenna for 5G applications. IEEE Transactions on Antennas and Propagation, 68(2), 1128–1133.
[23] Kumar, S., Dixit, A. S., Malekar, R. R., Raut, H. D., & Shevada, L. K. (2020). Fifth generation antennas: A comprehensive review of design and performance enhancement techniques. IEEE Access, 8, 163568–163593.
[24] Sun, L., Feng, H., Li, Y., & Zhang, Z. (2018). Compact 5G MIMO mobile phone antennas with tightly arranged orthogonal-mode pairs. IEEE Transactions on Antennas and Propagation, 66(11), 6364–6369.
[25] Wang, Y., & Du, Z. (2013). A wideband printed dual-antenna system with a novel neutralization line for mobile terminals. IEEE Antennas and Wireless Propagation Letters, 12, 1428–1431.

[26] Sun, L., Feng, H., Li, Y., & Zhang, Z. (2018). Tightly arranged orthogonal mode antenna for 5G MIMO mobile terminal. Microwave and Optical Technology Letters, 60(7), 1751–1756.
[27] Shi, X., Zhang, M., Xu, S., Liu, D., Wen, H., & Wang, J. (2017, March). Dual-band 8-element MIMO antenna with short neutral line for 5G mobile handset. In 2017 11th European Conference on Antennas and Propagation (pp. 3140–3142). IEEE.
[28] Nawaz, H., & Tekin, I. (2017). Double-differential-fed, dual-polarized patch antenna with 90 dB interport RF isolation for a 2.4 GHz in-band full-duplex transceiver. IEEE Antennas and Wireless Propagation Letters, 17(2), 287–290.
[29] Nawaz, H., & Tekin, I. (2017). Compact dual-polarised microstrip patch antenna with high interport isolation for 2.5 GHz in-band full-duplex wireless applications. IET Microwaves, Antennas and Propagation, 11(7), 976–981.
[30] Choi, J. I., Jain, M., Srinivasan, K., Levis, P., & Katti, S. (2010, September). Achieving single channel, full duplex wireless communication. In Proceedings of the Sixteenth Annual International Conference on Mobile Computing and Networking (pp. 1–12). Association for Computing Machinery.
[31] Debaillie, B., van den Broek, D. J., Lavin, C., Van Liempd, B., Klumperink, E. A., Palacios, C., ..., & Nauta, B. (2015, May). RF self-interference reduction techniques for compact full duplex radios. In 2015 IEEE 81st Vehicular Technology Conference (pp. 1–6). IEEE.
[32] Iwamoto, K., Heino, M., Haneda, K., & Morikawa, H. (2018). Design of an antenna decoupling structure for an inband full-duplex collinear dipole array. IEEE Transactions on Antennas and Propagation, 66(7), 3763–3768.
[33] Takahashi, H., Hirata, A., Takeuchi, J., Kukutsu, N., Kosugi, T., & Murata, K. (2012, October). 120-GHz-band 20-Gbit/s transmitter and receiver MMICs using quadrature phase shift keying. In 2012 7th European Microwave Integrated Circuit Conference (pp. 313–316). IEEE.
[34] Nagatsuma, T., Oogimoto, K., Yasuda, Y., Fujita, Y., Inubushi, Y., Hisatake, S., ..., & Lopez, G. C. (2016, September). 300-GHz-band wireless transmission at 50 Gbit/s over 100 meters. In 2016 41st International Conference on Infrared, Millimeter, and Terahertz Waves (pp. 1–2). IEEE.
[35] Shu, C., Hu, S., Yao, Y., & Chen, X. (2019, October). A high-gain antenna with dual circular polarization for W-band mm wave wireless communications. In 2019 Computing, Communications and IoT Applications (pp. 29–32). IEEE.
[36] Chizhik, D., Ling, J., & Valenzuela, R. A. (1998, October). The effect of electric field polarization on indoor propagation. In IEEE 1998 International Conference on Universal Personal Communications. Conference Proceedings (Vol. 1, pp. 459–462). IEEE.
[37] Lin, C. C., Kuo, L. C., & Chuang, H. R. (2006). A horizontally polarized omnidirectional printed antenna for WLAN applications. IEEE Transactions on Antennas and Propagation, 54(11), 3551–3556.
[38] Quan, X. L., Li, R. L., Wang, J. Y., & Cui, Y. H. (2012). Development of a broadband horizontally polarized omnidirectional planar antenna and its array for base stations. Progress in Electromagnetics Research, 128, 441–456.

[39] Sun, L., Li, Y., Zhang, Z., & Feng, Z. (2019). Compact co-horizontally polarized full-duplex antenna with omnidirectional patterns. IEEE Antennas and Wireless Propagation Letters, 18(6), 1154–1158.
[40] Makar, G., Tran, N., & Karacolak, T. (2016). A high-isolation monopole array with ring hybrid feeding structure for in-band full-duplex systems. IEEE Antennas and Wireless Propagation Letters, 16, 356–359.
[41] Kumari, K., Jaiswal, R. K., & Srivastava, K. V. (2019, December). A dual band full-duplex monopole antenna for WLAN application. In 2019 IEEE Indian Conference on Antennas and Propagation (pp. 1–4). IEEE.
[42] Rajesh Singh, Anita Gehlot, & Vishal Jain. (2019, September). Handbook of Research on the Internet of Things Applications in Robotics and Automation. DOI: 10.4018/978-1-5225-9574-8
[43] Shende, N., Gurbuz, O., & Erkip, E. (2013, March). Half-duplex or full-duplex relaying: A capacity analysis under self-interference. In 2013 47th Annual Conference on Information Sciences and Systems (CISS) (pp. 1–6). IEEE.
[44] Yang, M., Jeon, S. W., & Kim, D. K. (2017). Degrees of freedom of full-duplex cellular networks with reconfigurable antennas at base station. IEEE Transactions on Wireless Communications, 16(4), 2314–2326.
[45] Liu, G., Yu, F. R., Ji, H., Leung, V. C., & Li, X. (2015). In-band full-duplex relaying: A survey, research issues and challenges. IEEE Communications Surveys and Tutorials, 17(2), 500–524.
[46] Bharadia, D., McMilin, E., & Katti, S. (2013, August). Full duplex radios. In Proceedings of the ACM SIGCOMM 2013 conference on SIGCOMM (pp. 375–386). Association for Computing Machinery.
[47] Wójcik, D., Surma, M., Noga, A., & Magnuski, M. (2020). High port-to-port isolation dual-polarized antenna array dedicated for full-duplex base stations. IEEE Antennas and Wireless Propagation Letters, 19(7), 1098–1102.
[48] Amjadi, S. M., & Sarabandi, K. (2017). A low-profile, high-gain, and full-band subarray of cavity-backed slot antenna. IEEE Transactions on Antennas and Propagation, 65(7), 3456–3464.
[49] Amjadi, S. M., & Sarabandi, K. (2018, July). A compact, broadband, two-port slot antenna system for full-duplex applications. In 2018 IEEE International Symposium on Antennas and Propagation & USNC/URSI National Radio Science Meeting (pp. 389–390). IEEE.
[50] Nawaz, H., & Tekin, I. (2016, June). Dual polarized patch antenna with high inter-port isolation for 1GHz in-band full duplex applications. In 2016 IEEE International Symposium on Antennas and Propagation (pp. 2153–2154). IEEE.
[51] Miers, Z., Li, H., & Lau, B. K. (2013). Design of bandwidth-enhanced and multiband MIMO antennas using characteristic modes. IEEE Antennas and Wireless Propagation Letters, 12, 1696–1699.
[52] Foroozanfard, E., De Carvalho, E., & Pedersen, G. F. (2016, October). Decoupling of TX and RX antennas in a full-duplex mobile terminal. In 2016 International Symposium on Antennas and Propagation (pp. 236–237). IEEE.

Chapter 4

Comb-shaped microstrip patch antenna with defected ground structure for MIMO applications

Balwant Raj, Yadwinder Kumar, and Sunil Kumar

CONTENTS

DOI: 10.1201/9781003290230-4

4.1 INTRODUCTION

An antenna or aerial is used as a transition medium between the transmission line and a free space. A transmission line provides a path to transmit energy from the transmit end to the receiving end of the antenna. An antenna is the base design element in wireless communication systems. It can have different structures like a wire conducting design, an aperture design and an assembly of elements, or a patch to meet the requirements and specifications of a particular application. Over the last 50 years, various advances have been made in the field of antenna design, and with the revolution in communication technologies, antenna design and miniaturization have become very important. To overcome the limitations of bulky and nonconformal antennas, the microstrip patch antennas was developed in the 1970s [1]. It provides various advantages, and its outstanding performance has made it very popular for communication applications. Due to its inherent properties, the microstrip patch antenna became an important research topic in the past decade [2]. The antenna is suitable for commercial applications like mobile communication, satellites, and remote sensing systems and also for military applications. Microstrip patch radiating structures also have some restrictions like trivial bandwidth, reduced efficiency, modest gain, and surface waves [3]. To overcome these drawbacks, researchers developed new technologies for antenna design, and one of the best outcomes is the defected ground structure. This chapter gives a brief description of the microstrip patch antenna and the use of the defected ground structure to improve antenna performance.

4.1.1 Basic geometry of microstrip patch antennas

Presently, low-profile antennas are required for most commercial applications like mobile, radio, and wireless communications. These antennas are widely used for applications where performance constraints are size, weight, ease of installation, and cost such as satellite, spacecraft, mobile, missile, and aircraft applications. The microstrip patch radiating structure is based on modern printed circuit technology. Its geometry mainly consists of three layers—i.e., a conductive ground layer, a dielectric epoxy-like substrate, and a conductive patch layer [1].

4.1.2 Basic characteristics of microstrip patch antennas

A radiate patch is etched on the top of the substrate and on the back of the substrate. A metallic patch strip having a small thickness ($t \ll \lambda_0$) is placed at a small height ($h < \lambda_0$, usually $0.003\lambda_0 \leq h \leq 0.05\lambda_0$) above the ground plane. A dielectric substrate is used to separate the ground plane and patch. The radiating patch and ground layer are made up of a conducting material. Different substrate materials whose dielectric constant lies in the span of $2.2 \leq \varepsilon_r \leq 12$ like FR4 epoxy and Rogers material can be used in microstrip antennas. For better antenna performance, thick substrates that have lower dielectric constants are selected because they have large bandwidths and loosely bounded fields for space radiation and are highly efficient, but a thick substrate results in a larger antenna. For microwave applications, it is desirable to use thin substrates that have high dielectric constants because they require firmly bound fields for removing redundant radiation and result in smaller element sizes, but they have relatively smaller bandwidths and greater losses, which make them less efficient. Thus, the substrate material plays a significant part in calculating the measurements and bandwidth of an antenna [3, 4].

4.1.3 Various patch structures

A radiate patch is photoetched on the top of the substrate material. Usually, the patch length selected is about half of the dielectric wavelength, which corresponds to the resonant frequency. Different patch structures—e.g., rectangular, square, triangular, dipole, circular, ring shaped, and elliptical—may be used for microstrip antennas, but the commonly used are the rectangular, square, circular, and dipole shapes because of their ease of analysis and fabrication. Also, they provide excellent radiation characteristics such as low cross-polarization radiation. Dipole structures are used for arrays, as they provide larger bandwidths and require less space. Polarizations (both linear and circular) can be obtained by using either single elements or microstrip antenna arrays. The rectangular patch structure is the most widely used configuration for microstrip antenna designs as these antennas can be easily analyzed using the transmission line model or the cavity model, both of which give accurate results for thin substrates [4].

4.2 DIFFERENT FEEDING TECHNIQUES

A feed line is used to provide excitation or radiation to microstrip antennas. The various techniques for feeding microstrip patch antennas can be generally categorized as contact and noncontact. A brief comparison has

Table 4.1 Different feeding methods and their features [5]

Feature	*Coaxial probe*	*Microstrip Line*	*Aperture-coupled*	*Proximity-coupled*
Fabrication process	Easy (proper soldering and drilling requirement)	Easiest	Difficult (alignment requirement)	Difficult (alignment requirement)
Impedance matching	Simple	Simple	Simple	Simple
Spurious radiation	Low	High	Moderate	Low
Reliability	Poor	Best	Good	Good
Bandwidth	Narrow (2–5%)	Narrow (2–5%)	Narrow (2–5%)	Largest (13%)
Modeling	Difficult for thick substrate	Easy	Easy	Easy

been done in Table 4.1. Among these various techniques, the most popular and widely used are the coaxial feed, microstrip feed, aperture-coupled feed, and proximity-coupled feed. Microstrip line and coaxial probe are contact methods of feed, in which a direct contact is made and power is supplied directly to the patch via connecting elements like a microstrip line. Aperture coupling and proximity coupling are noncontact feed methods, in which power is transferred to the patch from the microstrip line using an electromagnetic field coupling.

4.2.1 Coaxial probe feeding

In this feeding method, the patch is connected to the inner conducting coaxial prove and the ground plane to the outer conducting coaxial probe.

Advantages of coaxial probe feeding

1. It removes spurious radiation.
2. It is easily fabricated.
3. Matching can be easily done.

Disadvantages of coaxial probe feeding

1. It results in narrowband width.
2. Modeling is difficult for thick substrate.
3. The method Possesses inherent asymmetries that result in cross-polarization radiation.

4.2.2 Microstrip line feeding

In this feeding method, a radiating patch has a direct connection with the conducting strip, which has a width much smaller than the patch width and is considered as an extension of the patch.

Advantages of microstrip line feeding

1. It is simple and easy to fabricate.
2. Modeling is easy, and matching is obtained by simply curbing the inset position.

Disadvantages of microstrip line feeding

1. Increased surface wave excitation and spurious feed radiation result from an increase in substrate thickness and cause limited bandwidth.
2. Cross-polarization occurs due to radiation.

4.2.3 Aperture-coupled feeding

This noncontact feeding method has been introduced to overcome the problems of contact feeding methods. Here, separation between the two substrates is provided by a ground plane. A microstrip feed line placed on the bottom side of the lower substrate provides energy coupling to the patch by introducing a slit-like opening in the ground surface. Generally, the bottom substrate is made up of a thin material with a high dielectric constant, whereas the top substrate is made up of a thick material with a low dielectric constant.

A ground is used to provide isolation between the radiating and feed elements and also to reduce interference due to spurious radiation in order to improve polarization and pattern formation. Feed-line width and slot length are designed facilitate matching and optimization.

Advantages of aperture-coupled feeding

1. The radiating element and feeding mechanism can be optimized independently.
2. Modeling is easy and minimizes spurious radiation.
3. There is polarization purity with no cross-polarization.

Disadvantages of aperture-coupled feeding

1. The fabrication process is the most difficult compared to other feeding methods.
2. The method suffers from limited bandwidth.

4.2.4 Proximity-coupled feeding

Proximity coupling is another name for electromagnetic coupling. In this feeding method, a microstrip feed is placed between both dielectric substrates and the patch element. It is engraved from the top side of the upper substrate. By simply curbing the width/length ratio of the patch and feeding the stub's length, matching can be achieved.

Advantages of proximity-coupled feeding

1. Bandwidth is the largest among the four feeding methods.
2. Modeling is easy and provides spurious-free radiations.

Disadvantages of proximity-coupled feeding

1. Substrate thickness increases.
2. Fabrication is difficult.

4.2.5 Calculation of microstrip patch antenna dimensions

- To compute the width (W) of the patch:

$$W = \frac{C}{2f_0\sqrt{\left(\frac{\epsilon_{r+1}}{2}\right)}} \tag{4.1}$$

- To compute the effective dielectric constant (ε_{eff}) *of* the substrate:

$$\epsilon_{eff} = \frac{\epsilon_{r+1}}{2} + \frac{\epsilon_{r-1}}{2}\left(1+12\frac{h}{w}\right)^{-\frac{1}{2}} \tag{4.2}$$

- To compute the effective length (L_{eff}) of the patch:

$$L_{ff} = \frac{c}{2f_0\sqrt{\epsilon_{eff}}} \tag{4.3}$$

- To compute the extension length (ΔL) of the patch:

$$\Delta L = 00.412h\frac{\left(\epsilon_{eff}+00.3\right)\left(\frac{w}{h}+00.264\right)}{\left(\epsilon_{eff}-00.258\right)\left(\frac{w}{h}+00.8\right)} \tag{4.4}$$

- To compute the actual length (L) of the patch:

$$L = L_{eff} - 2\Delta L \tag{4.5}$$

The parameters are as follows:

F_0 is the resonant frequency.
W is the width of the patch.
L is the length of the patch.

h is the thickness of the substrate.
ε_r is the relative permittivity of the substrate.
c is the speed of light: 3×10^8 m/s.

4.3 RADIO SPECTRUM

The radio spectrum covers a wide range of frequencies on the electromagnetic spectrum as shown in Table 4.2 [6].

4.4 ADVANTAGES AND DISADVANTAGES OF MICROSTRIP PATCH ANTENNAS

The microstrip patch antenna provides various advantages, including the following:

1. These antennas have a low profile, are lightweight, and have a planar configuration.
2. They are concordant to the surfaces.
3. They are easy to fabricate, as they utilize printed circuit technology, and the manufacturing cost is also low.
4. They are mechanically robust and can be easily mountable on stiff surfaces.
5. They are compatible with microwave integrated circuits.
6. Multifrequency operations are possible.

Table 4.2 Various radio frequency bands and their applications [6]

Band designation	*Frequency range (GHz)*	*Applications*
HF	0.003–0.03	Military and government communication system
VHF	0.03–0.3	FM radio and TV broadcasting
UHF	0.3–1.00	Wi-Fi, Bluetooth, medical telemetry
L	1–2	GSM mobile, medical telemetry
S	2–4	Satellite, microwave oven, biomedical
C	4–8	Ship navigation
X	8–12	Radar, weather monitoring, air traffic control
Ku	12–18	VSAT system on yachts and ships
K	18–27	GPS
Ka	27–40	HD satellite TV
V	40–75	Radar and other scientific research
W	75–110	Satellite communication, military radar targeting and tracking
mm or G	110–300	Scientific research and telecommunication

Microstrip patch antennas also have limitations such as the following:

1. They have smaller bandwidths.
2. They are less efficient.
3. They have small gain and power.
4. They cause spurious feed radiation.
5. Surface wave excitation degrades the antenna performance.

4.5 DEFECTED GROUND STRUCTURE

Besides having various advantages, conventional microstrip patch antennas suffer from some major drawbacks like low gain, narrow bandwidth, large size, single operating frequency, surface wave excitation, and cross polarization that limits their performance. To overcome these limitations and enhance the performance parameters of these antennas, various new technologies have emerged like stacking, the photonic bandgap (PBG), the electromagnetic bandgap (EBG), and metamaterials [7]. The defected ground structure (DGS) is one of the new concepts that has been applied in the field of microwave circuits over the past few years. A brief comparison between DGS and PBG has been done in Table 4.3. The microstrip patch antenna with a DGS has gained much attention among all the proposed techniques for improving antenna performance, as it provides a simple structural antenna design and is less complex. A PBG structure that utilizes a metallic ground plane has been proposed in [8]. The PBG has a periodic structure and is used to reject certain frequencies. However, for micro and millimeter-wave components, modeling of the PBG structure is not easy, which limits the demand for these components and circuit designs. Also, there are limiting factors that affect the performance of the PBG like spurious radiation due to etched periodic defects, lattice numbers, shapes, and spacing [7].

Table 4.3 Comparison between PBG and DGS [15]

Characteristic	*PBG*	*DGS*
Structure	Periodic etched unit	One or a few etched units
Circuit area	Large	Relatively small
Equivalent circuit Extraction	Difficult	Easy
Precision	Low	High
Spurious radiation	Large	Less
Modeling	Difficult	Easy
Microwave circuit Properties	Similar	Similar
Fabrication	Difficult	Easy

Likewise, another technique proposed is the ground plane aperture (GPA), which comprises the microstrip line with a slot at the center in the ground plane. The GPA has various attractive applications ranging from filters for removing spurious bands to 3 dB edge couplers for tight coupling. Introducing the GPA below the strips results in a significant change in the line properties due to the variation in the characteristic impedance width of the GPA. To address these limitations, a DGS that exhibits a band-stop property is proposed in [9]. Researchers are exploring various novel DGSs due to their wide range of applications in microwave antenna design. This chapter includes a detailed study of the DGS.

4.5.1 Why a DGS is needed?

The microstrip patch antenna has degraded circuit performance due to its various disadvantages like small bandwidth, low gain, less efficiency, and surface wave excitation. Bandwidth boosting is necessary for most of the practical applications to fulfill the demand of wireless communication, as the numbers of users is growing day by day. There are different techniques that can be used to increase the bandwidth of patch antennas: e.g., increasing the height of the patch or the thickness of the substrate and decreasing the permittivity of the substrate [10]. With the advancement in the technology, researchers are trying to miniaturize the microstrip patch antenna and make it as compact as possible. Thus, to increase the bandwidth of the microstrip antenna while keeping the antenna small, the DGS technique was proposed [11, 12]. The DGS also removes the surface waves and rejects certain frequencies [13, 14].

4.5.2 DGS structure

The DGS technique, as its name implies, introduces a defect that can be regular or irregular on the ground plane of the transmission line. An engraving defect on the ground disturbs the shielded current distribution in the ground plane. Transmission line characteristics such as line inductance and capacitance will also change due to the current disturbance. Thus, any defect etched on the ground increases the effective inductance and capacitance of the transmission line.

The basic DGS has a dumbbell shape and therefore is often called a dumbbell DGS [9]. It comprises two rectangular defected areas of $a \times b$ and a narrow connecting slot with a $g \times w$ gap in the lower ground layer.

1. *Various DGS units*

   Various simple and complex DGSs that are etched on the ground plane have been reported in the literature. Some of the defect geometries have a spiral head shape, arrowhead slot, or H-shaped slot, and others have much more complex shapes like a middle slotted square open-loop, dumbbell with open loop, and interdigital DGS to improve circuit performance. These novel proposed designs have many advantages over the dumbbell DGS like more compact designs, a higher slow-wave effect, and increased bandwidth.

2. *Periodic DGS*
   Periodic structures etched in the transmission line have wide applications in microwave antenna design and have received much interest over the past few years. These structures provide finite passband and stopband characteristics. Periodic structures have additional equivalent components and higher slow-wave effects, which are utilized to reduce antenna size. In a periodic DGS, the physical structure is repeated or cascaded in the ground plane. It restricts the stopband attributes for the proposed antenna depending on the number of periodic DGS cells introduced on the ground. Proper care must be taken in selecting the structure of and distance between two DGS units. Periodic DGSs can be classified into two types: the horizontal periodic DGS (HPDGS) and the vertical periodic DGS (VPDGS).
   a. *Horizontal periodic DGS*
      In a HPDGS, periodic structures are cascaded serially along the microstrip transmission line. Conventionally, the HPDGS is used for planar microstrip lines, which aim to enlarge the stopband response.
   b. *Vertical periodic DGS*
      In a VPDGS, periodic structures are arranged vertically. The VPDGS provides a higher slow-wave effect compared to the HPDGS. For the same physical dimensions, an increased slow-wave effect implies greater electrical length. Circuit size is reduced in the VPDGS.

4.5.3 DGS transmission characteristics

Filtering unwanted signals and tuning to higher-order harmonics can be easily achieved by etching the required defect slot in the ground plane to meet the circuit requirements without making the circuit complex. The stopband effect, slow-wave effect, and high impedance are the characteristics of the DGS [12].

1. *Stopband effect*
   Engraving different DGSs (periodic or nonperiodic) in the metallic ground layer provides rejection of certain frequency bands, which is known as the stopband effect of a DGS. This characteristic of the DGS is utilized to suppress surface waves and provides transmission free of spurious radiation and leakage. This frequency-selective property of the DGS has been widely used in microwave filter applications and

has gained much interest as a research hotspot. The DGS improves circuit performance, as it provides a sharp frequency transition, higher harmonic suppression, a wider stopband response, and excellent passband and stopband characteristics by removing ripples.

2. *Slow-wave effect*
   This very important characteristic of the DGS is induced by the LC equivalent circuit. A transmission line with etched DGS has very high impedance and an increased slow-wave effect compared to a conventional line without the DGS. These properties are utilized to reduce antenna size. The performance of the antenna improves by using the DGS, which suppresses the surface waves and limits the mutual coupling of the antenna.
3. *High impedance*
   Microstrip line impedance can increased to a value greater than 200 Ω by introducing a DGS in the metallic ground plane, which results in an increase in the equivalent inductance L and a decrease in the capacitance C. The conventional microstrip line has limited acceptable impedance in the range of 100 to 130 Ω. This property of the DGS is used for many applications like digital systems.

4.5.4 DGS equivalent circuit

The metallic part of the microstrip antenna is a combination of distributed resistance, capacitance, and inductance. Babinate's principle states that each slot is reciprocal to its metallic structure. For practical realization of the proposed DGS unit, one of the requirements is to elicit the equivalent circuit components. The parameters of the circuit can be extracted from the equivalent circuit design using a simulation result (scattering parameter vs. frequency) obtained by a full-wave electromagnetic simulator to properly identify the cutoff frequency and low-pass response. The DGS equivalent circuit consists of a parallel tuned circuit in series with the transmission line to be coupled [8]. DGS equivalent circuits can be classified into three types:

1. LC and RLC equivalent circuit.
2. Π-shaped equivalent circuit.
3. Quasistatic equivalent circuit.

The most efficient equivalent models of the DGS unit for transmission lines have parallel R, L, and C resonant circuits. The dumbbell DGS, composed of two rectangular defected areas connected by a narrow slot, can be considered as an equivalent L and C circuit. The rectangular parts are responsible for increasing the current route length and the effective inductance. The slot part is responsible for increasing the effective capacitance due to the accumulation of a charge. The parallel

LC circuit corresponds to the resonant frequency [16]. With an increase in the etched rectangular area, there is an increase in the effective series inductance that corresponds to the lower cutoff frequency. Moreover, with the increase in induced gap distance, there is a decrease in the effective capacitance, which moves the attenuation poles to a higher frequency. The resistance part is responsible for conduction, radiation, and the dielectric losses in the defect. The DGS can be matched to the Butterworth low-pass filter by equalizing the reactance values of both circuits at the cutoff frequency [7].

The major disadvantage is that direct correlation is not found between the DGS physical dimensions and the equivalent LC parameters. It is not possible to predict the performance of the DGS until the optimized results are obtained by the trial-and-error method.

4.5.5 Advantages of a DGS

The DGS has gained much importance in the research field, as it offers various advantages and has wide application areas in microwave antenna design. Some of the advantages are as follows:

1. It provides higher operating bandwidth and reduced antenna size.
2. It provides less return loss and high gain.
3. It makes the system compact and effective.
4. It has frequency-selective properties that are utilized to reject undesired frequencies.
5. It suppresses surface waves and provides arbitrary stopbands.
6. It is very suitable for wireless applications.

4.6 ANTENNA DESIGN

4.6.1 Basic rectangular patch

The dimensions (length and width) of the rectangular patch have been calculated using Equations 4.1–4.5 and a resonant frequency of 1.5 GHz and are found to be 21 mm and 21 mm. respectively. The 0th iteration of the antenna design is shown in Figure 4.1.

4.6.1.1 First iteration of the rectangular patch, introducing comb-shaped slots

The first iteration of the proposed antenna introduces comb-shaped fractal slots on opposite sides of the rectangular patch, as shown in Figure 4.2; other dimensions are as for the 0th iteration. The parametric values of the comb-shaped fractal slots are given in Table 4.4.

Figure 4.1 Conventional rectangular patch [17].

Figure 4.2 First iteration of the rectangular patch [18].

Table 4.4 Parametric values of the comb-shaped fractal slots in the first iteration

S. no.	*Parameter*	*Value*
1	Length of comb-shaped slot	7 mm
2	Width of comb-shaped slot	3 mm

Table 4.5 Parametric values of the comb-shaped fractal slots in the second iteration

S. no.	*Parameter*	*Value*
1	Length of comb-shaped slot	2.33 mm
2	Width of comb-shaped slot	0.43 mm

4.6.1.2 *Second iteration of the rectangular patch with comb-shaped slots*

With the first iteration as the base geometry, the fractal slots are repeated in the second iteration to create more fractals, the parametric values for the second iteration are given in Table 4.5.

Details of the comb structure and various dimensions of the proposed antenna geometry are shown in Figure 4.3. It also shows the dimensions of the enlarged portion which otherwise is not easy to judge by naked eyes. Obtained geometric values of the proposed antenna are shown in Table 4.6.

4.6.2 DGS ground layout

A DGS has been used in the ground plane of the suggested antenna. A G-shaped pattern is used as the DGS ground, which disorganizes current distribution in the ground plane. The suggested G-shaped DGS has a slot width of 3 mm. The complete geometry is shown in Figure 4.4 and its geometric values are shown in Table 4.7.

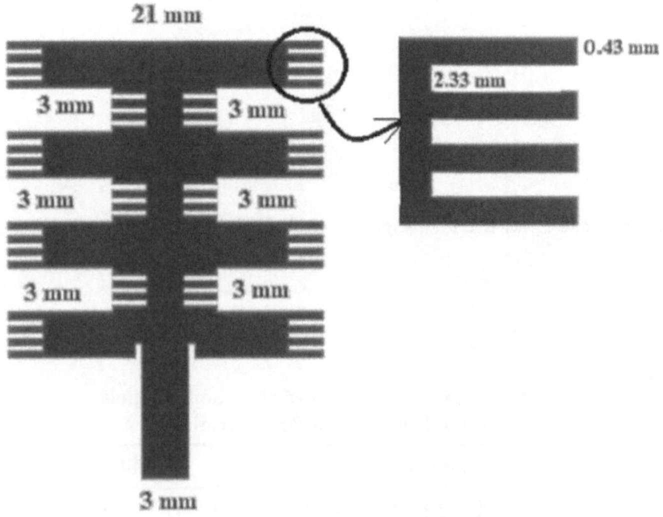

Figure 4.3 Geometrical details of suggested comb-shaped radiator layout.

Table 4.6 Geometrical values of the proposed antenna

Antenna design parameter	*Description*
Length	35 mm
Width	35 mm
Substrate material	FR4 epoxy
Dielectric constant	$\varepsilon_r = 4.4$
Thickness	1.2 mm
Length and width of patch	21 mm
Feed length	10 mm
Feed width	3 mm

Figure 4.4 G-shaped DGS ground layout.

4.7 SIMULATED RESULTS

This section introduces various antenna parameters such as return loss, voltage standing wave ration (VSWR), radiation characteristics, and gain that are caused by iterations on the patch and the introduction of a DGS in the ground plane.

Table 4.7 Geometrical values of the DGS ground

Parameter	*Description*
Length of ground	35 mm
Width of ground	35 mm
Inner width	3 mm
Dielectric constant	$\varepsilon_r = 4.4$
Thickness of substrate	1.2 mm

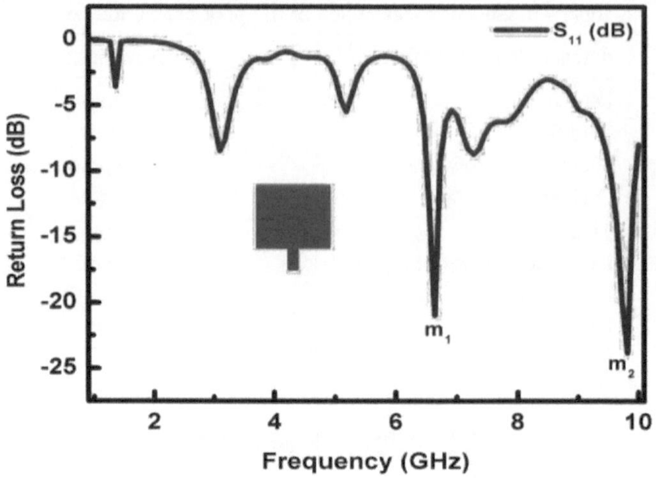

Figure 4.5 Return loss in the simple rectangular patch.

4.7.1 Return loss of proposed antenna structure

The graph in Figure 4.5 determines that the reflection coefficients of the proposed rectangular patch antenna are –20.9861 dB for 6.6364 GHz and –23.8244 dB for 9.8182 GHz. Figures 4.5–4.7 show the return loss versus frequency graphs for various iterations of the suggested comb-shaped pre-fractal antenna. It is clearly seen that with increasing iterations, the return

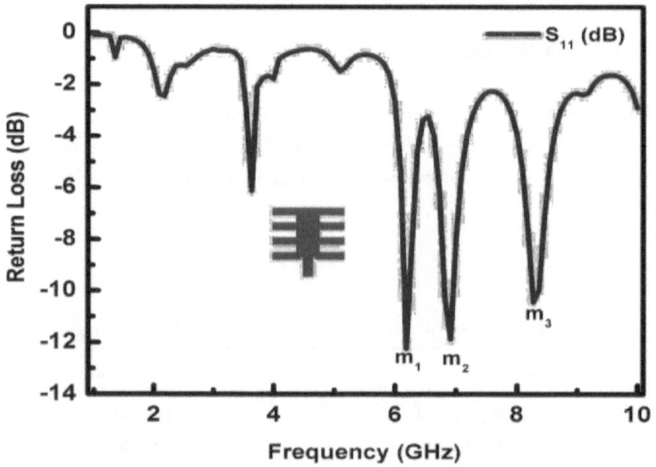

Figure 4.6 Return loss in the first iteration of the patch, introducing comb-shaped slots.

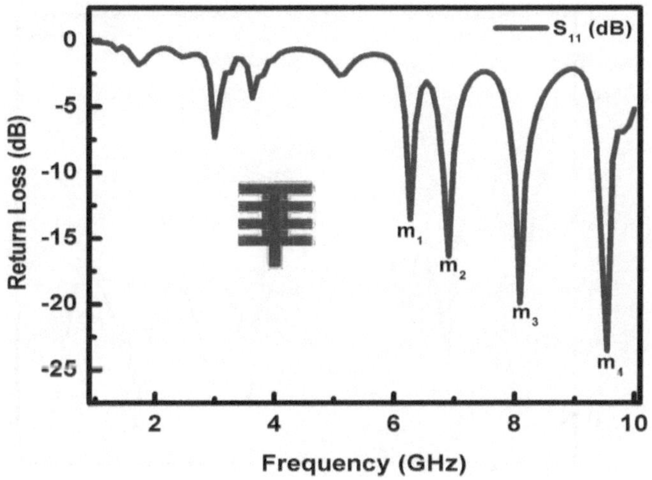

Figure 4.7 Return loss in the second iteration of the rectangular patch, introducing more comb-shaped slots.

loss and resonant frequency bands increase. The acceptable values of return loss are less than –10 dB. The values of the return loss at different frequencies for three iterations of the proposed antenna are at the acceptable level, as shown in Table 4.8.

4.7.2 Voltage standing wave ratio

The VSWR indicates the capability of the radiating structure to match its impedance with that of the transmission line or cable. Its value for a perfectly matched antenna should be less than 2. VSWR values obtained from

Table 4.8 Details of different resonant frequencies for different iterations

Iteration	*Peak of resonant curves*	*X (Frequency, GHz)*	*Y (Return Loss, dB)*
0	m_1	6.6364	–20.9861
	m_2	9.8182	–23.8244
1	m_1	6.1818	–12.2298
	m_2	6.9091	–11.8693
	m_3	8.2727	–10.4349
2	m_1	6.2727	–13.5127
	m_2	6.9091	–16.3283
	m_3	8.0909	–19.8724
	m_4	9.5455	–23.5253

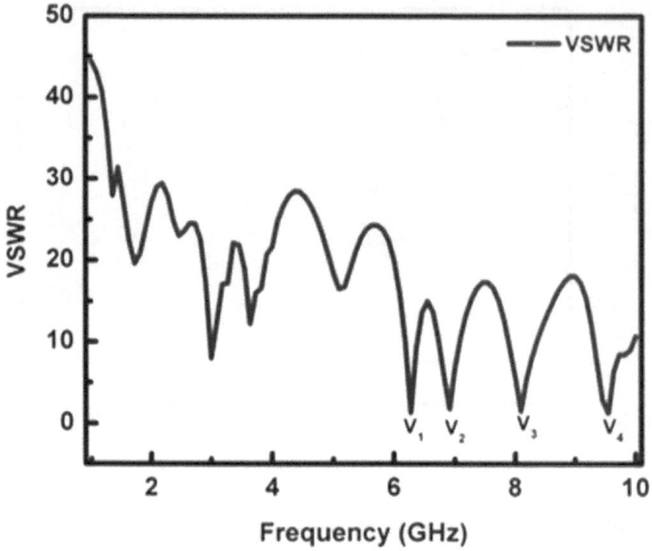

Figure 4.8 VSWR of the suggested comb-shaped antenna with a DGS ground.

the simulation of the suggested comb-shaped antenna with a G-shaped DGS ground are 1.3407 for 6.2727 GHz, 1.8045 for 6.9091 GHz, 1.5667 for 8.0909 GHz, and 1.3236 for 9.5455 GHz. All the values are in acceptable range—i.e., less than 2—as shown in Figure 4.8.

4.7.3 Comparison of various iterations

Reflection coefficients for various iterations are compared in the single graph in Figure 4.9. It clearly demonstrates that the number of resonant frequencies increases as the iterations increase. For a simple rectangular patch, we have obtained resonances at two frequencies; for the first iteration, we obtained three resonant frequencies; and, finally, for the third iteration, we obtained four resonant frequencies.

4.7.4 Current distribution

The current distributions at the resonating frequencies were evaluated and are presented in Figure 4.10. The maximum current accumulates at the patch and feed line at the lowest resonating frequency. The current starts flowing at the lower half of the patch and contributes to the resonating frequencies of 6.27 GHz, 6.90 GHz, 8.09 GHz, and 9.54 GHz. As the resonating frequency is higher at 9.54 GHz, the current is concentrated in the full patch. The current flows in the outer perimeter of the patch and results in the higher resonating frequency [19].

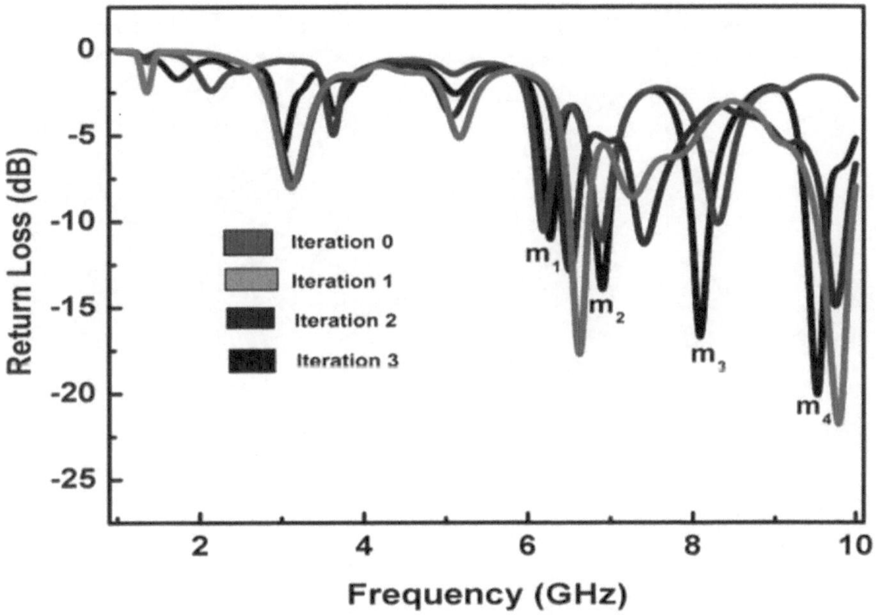

Figure 4.9 Comparison of return loss values for various iterations.

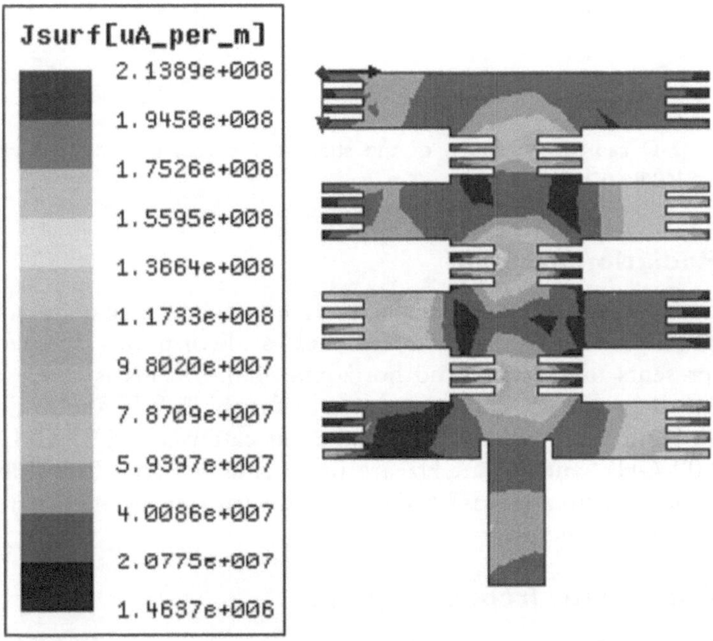

Figure 4.10 Current distribution at 6.27 GHz.

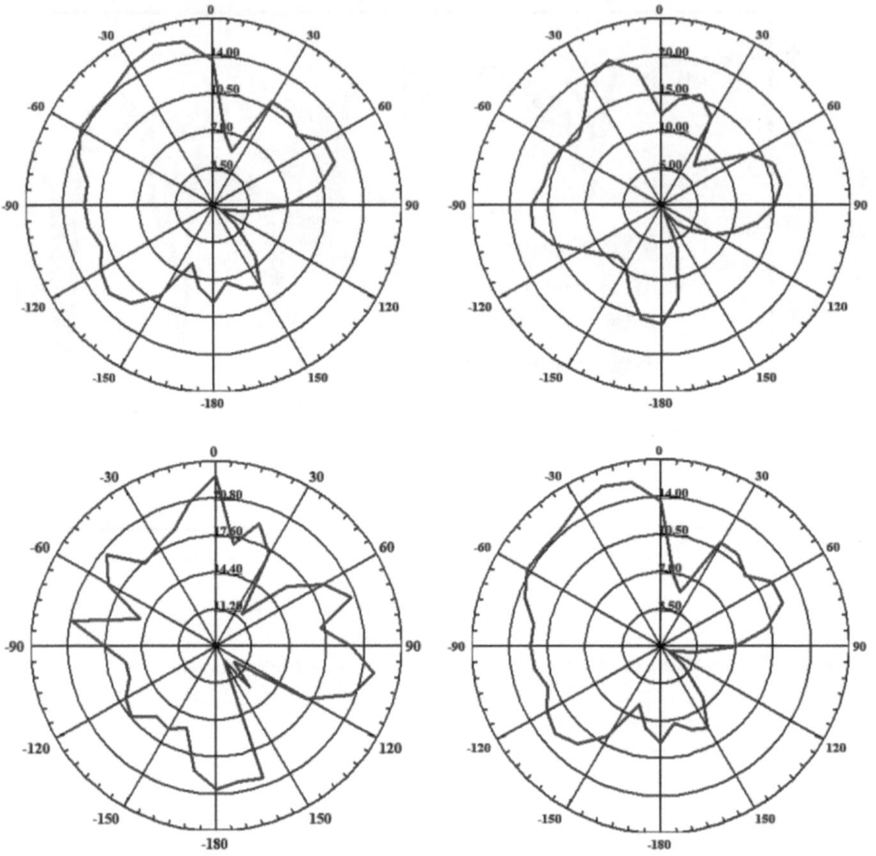

Figure 4.11 2-D radiation pattern of the suggested antenna at all four resonant frequencies.

4.7.5 Radiation pattern

A two-dimensional graph represents the radiation property of an antenna. The plotted quantities are shown in a linear scale or in dB. The plot typically represents the vertical and horizontal planes. This is often known as a polar diagram. The radiation patterns for the E-field and H-field are shown in Figure 4.11. The E-plane radiation patterns at 6.27 GHz, 6.90 GHz, 8.09 GHz. and 9.54 GHz are in acceptable form [20]. Table 4.9 shows values of gain and VSWR related to the various resonating frequencies for the final structure.

4.7.6 Gain versus frequency graph

Gain values at all the resonant frequencies are shown in Table 4.9. The values of antenna gain are 3.41, 2.35, 3.04, and 3.74, respectively, at the

Table 4.9 Values of gain and VSWR corresponding to the resonating frequencies for the final structure of the proposed antenna structure

S. no.	*Resonant frequency (GHz)*	*Return loss (S_{11}) (in dB)*	*VSWR*	*Gain*
1	6.27	−13.51	1.34	3.41
2	6.90	−16.32	1.80	2.35
3	8.09	−19.87	1.56	3.04
4	9.54	−23.52	1.32	3.74

four different resonant frequencies: 6.27 GHz, 6.90 GHz, 8.09 GHz, and 9.54 GHz. The values obtained are in good agreement and within the acceptable range. Figure 4.12 shows the graph of gain versus frequency. The slight differences in antenna gains can be attributed to the effects of conductor and dielectric loss [21, 22].

For an efficient transfer of energy, the impedances of the antenna and the transmission cable must be the same [14]. Transceivers and their transmission lines are typically designed for 50 Ω impedance [22]. If the antenna has an impedance different from 50 Ω, then there is a mismatch, and an impedance-matching circuit is required [23]. The proposed antenna has good impedance-matching characteristics.

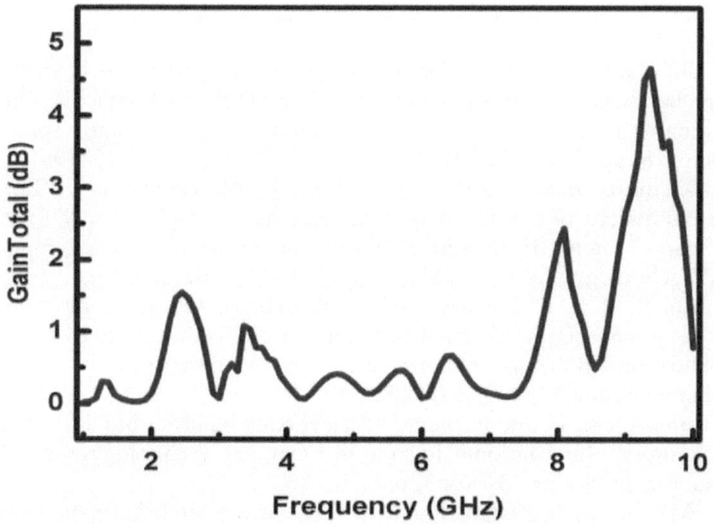

Figure 4.12 Gain versus frequency graph for the proposed antenna.

4.8 CONCLUSION

This chapter has presented a brief discussion of the DGS and the design and analysis of a comb-shaped microstrip antenna using a DGS. To improve the performance of a simple microstrip antenna, various iterations and a G-shaped DGS are used. In order to achieve a multiband operation, a G slot has been etched in the ground plane of this antenna, altering the overall current distribution on the ground plane. The return losses of the proposed multiband antenna at different frequencies are –13.51 dB (6.27 GHz), –16.32 dB (6.90 GHz), –19.87 dB (8.09 GHz), and –23.8244 dB (9.54 GHz). The satisfactory VSWR values for all resonating frequencies confirm good impedance matching. Gain varies between 2.35 dB and 3.74 dB for all the resonant frequencies, with maximum gain at 9.54 GHz. The antenna prototype has been simulated, and the measured $S_{11,}$ VSWR, gain, and other results like current distribution, impedance matching, and directivity are in good agreement. Moreover, the proposed antenna is very compact (21 mm × 21 mm × 1.2 mm), low in cost (using easily available and economical materials) and very simple to design, and it shows very good radiation characteristics over its operating frequency bands. The multiplexing efficiency, peak gain, and defected ground (DG) show that the proposed antenna can be used for various multiple-input multiple-output applications. Multiband antennas can also be used for multiple wireless communication technologies such as Wi-Fi, wireless LAN, and LTE.

REFERENCES

[1] C. A. Balanis, Antenna Theory; Analysis and Design, 3rd ed. Hoboken, NJ: Wiley, 2005.

[2] Abedin, Zain & Ullah, Dr. Zahid, "Design of a Microstrip Patch Antenna with High Bandwidth and High Gain for UWB and Different Wireless Applications" International Journal of Advanced Computer Science and Applications. Volume 8, pp. 379-382, 2017 DOI: 10.14569/IJACSA.2017.081049.

[3] M. Khaliluzzaman, D. K. Chy, and M. R. H. Chowdhury, "Enhancing the Bandwidth of a Microstrip Patch Antenna at 4 GHz for WLAN Using H-Shaped Patch," International Journal of Innovative Science, Engineering and Technology, Volume 2, No. 2, pp. 231–237, 2015.

[4] V. Mokal, P. S. R. Gagare, and R. P. Labade, "Analysis of Micro Strip Patch Antenna Using Coaxial Feed and Micro Strip Line Feed for Wireless Application," IOSR Journal of Electronics and Communication Engineering, Volume 12, No. 3, pp. 36–41, 2017.

[5] A. Kumar, J. Kaur, and R. Singh, "Performance Analysis of Different Feeding Techniques," International Journal of Emerging Technology and Advanced Engineering, Volume 3, No. 3, pp. 884–890, 2013.

[6] M. M. Dawoud, "High Frequency Radiation and Human Exposure," Proceedings of the International Conference on Non-ionizing Radiation at UNITEN: Electromagnetic Fields and Our Health, pp. 1–7, October 2003.

[7] L. H. Weng, Y. C. Guo, X. W. Shi, and X. Q. Chen, "An Overview on Defected Ground Structure," Progress in Electromagnetics Research B, Volume 7, pp. 173–189, 2008.

[8] E. Yablonovitch, "Inhibited Spontaneous Emission in Solid-State Physics and Electronics," IET Conference Publications, Volume 58, p. 20, 1987.

[9] J. I. Park et al., "Modeling of a Photonic Bandgap and Its Application for the Low-Pass Filter Design," Asia-Pacific Microwave Conference Proceedings, Volume 2, pp. 331–334, 1999.

[10] A. Khanna, D. K. Srivastava, and J. P. Saini, "Bandwidth Enhancement of Modified Square Fractal Microstrip Patch Antenna Using Gap-Coupling," Engineering Science and Technology, Volume 18, No. 2, pp. 286–293, 2015.

[11] P. A. Nawale and R. G. Zope, "Design and Improvement of Microstrip Patch Antenna Parameters Using Defected Ground Structure," Journal of Engineering Research and Applications, Volume 4, No. 3, pp. 123–129, 2014.

[12] M. K. Khandelwal, B. K. Kanaujia, and S. Kumar, "Defected Ground Structure: Fundamentals, Analysis, and Applications in Modern Wireless Trends," International Journal of Antennas and Propagation, Volume 2017, No. 1, pp. 1–22, 2017.

[13] M. D. S. Salgare and S. R. Mahadik, "A Review of Defected Ground Structure for Microstrip Antennas," International Research Journal of Engineering and Technology, Volume 2, No. 6, pp. 150–154, 2015.

[14] A. K. Arya, M. V. Kartikeyan, and A. Patnaik, "Defected Ground Structure in the Perspective of Microstrip Antennas: A Review," Frequency, Volume 64, No. 5–6, pp. 79–84, 2010.

[15] V. S. Melkeri, M. Lakshetty, and P. V Hunagund, "Microstrip Antenna with Defected Ground Structure: A Review," International Journal of Electrical Electronics and Telecommunication Engineering, Volume 46, No. 1, pp. 2051–3240, 1492.

[16] B. D. Orban and G. J. K. Moernaut, "The Basics of Patch Antennas," Orban Microwave Products, 2005. https://www.rfglobalnet.com/doc/technical-article-the-basics-of-patch-antenna-0001

[17] A. Nagpal, S. S. Dillon, and A. Marwaha, "Multiband E-Shaped Fractal Microstrip Patch Antenna with DGS for Wireless Applications," in 2013 5th International Conference on Computational Intelligence and Communication Networks, pp. 22–26, 2013.

[18] B. D. Patel, T. Narang, and S. Jain, "Microstrip Patch Antenna—A Historical Perspective of the Development," Conference on Advances in Communication and Control Systems, pp. 445–449, 2013.

[19] D. Upadhyay and R. P. Dwivedi, "Antenna Miniaturization Techniques for Wireless Applications," IFIP International Conference on Wireless and Optical Communications Networks, pp. 1–4, 2014.

[20] N. S. Dandgavhal, M. B. Kadu, and R. P. Labade, "Bandwidth and Gain Enhancement of Multiband Fractal Antenna Using Suspended Technique," European Journal of Advances in Engineering and Technology, Volume 2, No. 7, pp. 38–42, 2015.

[21] B. Kohitha, "Pentagonal Shaped Microstrip Patch Antenna in Wireless Capsule Endoscopy System," Computer Science Conference Proceedings, Volume 4, pp. 47–54, Jan. 2012.

[22] V. Kaushal, T. Singh, V. Kumar, and A. Kumar, "A Comprehensive Study of Antenna Terminology Using HFSS," International Journal on Electronics and Communication Technology, Volume 5, No. Spl-1, pp. 24–29, 2014.
[23] Y. Kumar and S. Singh, "Performance Analysis of Coaxial Probe Fed Modified Sierpinski–Meander Hybrid Fractal Heptaband Antenna for Future Wireless Communication Networks," Wireless Personal Communications, Volume 94, No. 4, pp. 3251–3263, 2017.

Chapter 5

Multiband MIMO antennas

Harsh Verdhan Singh, D. Venkata Siva Prasad, and Shrivishal Tripathi

CONTENTS

DOI: 10.1201/9781003290230-5

5.1 INTRODUCTION

Today practically uninterrupted connectivity is required for wireless devices, and the size of these devices is continuously decreasing, resulting in the consistent requirement for a comparable size reduction in the antenna elements. Despite this, the performance parameters of the antenna are still unchanged and in fact are getting more complex every day with this size reduction. As the demands of advanced wireless communication emerge, the performance requirements for antennas becomes more complex and more challenging to achieve.

In the past decade, communication with wireless devices has dramatically developed. During the initial utilization of the cellular, personal communications system (PCS), digital cellular system (DCS), and Global System for Mobile Communications (GSM) networks, the mobile device usually had to function within a specified band, designated by the explicit carrier's license(s). In the present situation, the mobile device frequently has to function in multiple bands, which could comprise several 802.11 (Wireless Fidelity or Wi-Fi), GSM, Global Positioning System (GPS), and 802.16 (Worldwide Interoperability for Microwave Access or WiMAX) frequencies, as defined in Table 5.1. The trend is for mobile devices to carry out numerous tasks: for example, wireless network access (through personal, local, metropolitan, and wide area networks—PAN, LAN, MAN, and WAN, respectively), mobile TV access (DVB-NGH), navigation (using global navigation systems or GNSs), radio-frequency identification (RFID), and contactless payment. Therefore, the most important enhancement in hardware, in an effort to offer global roaming functionality and signal diversity, is the multiple-input multiple-output (MIMO) antenna with the multiple antenna links required for multiband and multistandard tasks [1]. Technologies using present and future communication standards such as Long-Term Evolution (LTE), Universal Mobile Telecommunications Service (UMTS), GSM, wireless local area network (WLAN), WiMAX, and near-field communication (NFC) prefer to be available on the same device rather than having the user maintain several devices [2, 3]. The spectrum distribution of the different classified services usually used in existing mobile and handheld devices is shown in Table 5.1 [4–13]. From the antenna engineer's perspective, an antenna that functions uninterruptedly without tuning from 380 MHz through 2,500 MHz is designated as a single band but also as a broadband antenna. In today's wireless device terminology, an antenna

Table 5.1 Important wireless communications technology and frequency bands [4–13]

Technology	f_0 *(GHz)*	*Uplink (GHz)*	*Downlink (GHz)*	*Channel number*
GSM-450 [4]	0.45	0.4506–0.4576	0.4606–0.4676	259–293
GSM-480 [4]	0.48	0.479–0.486.0	0.489–0.496	306–340
T-GSM-380 [4]	0.38	0.3802–0. 3898	0.3902–0.3998	Dynamic
T-GSM-410 [4]	0.41	0.4102–0.4198	0.4202–0.4298	Dynamic
GSM-710 [4]	0.71	0.6982–0.7162	0.7282–0.7462	Dynamic
GSM-750 [4]	0.75	0.7772–0.7922	0.7472–0.7622	438–511
GSM-850 [4]	0.85	0.8242–0.848.8	0.8692–0.8938	128–251
T-GSM-810 [4]	0.81	0.8062–0.8212	0.8512–0.8662	Dynamic
P-GSM-900 [4]	0.90	0.890–0.915	0.935–0.960	1–124
E-GSM-900 [4]	0.90	0.880–0.915	0.925–0.960	975–1023, 0–124
R-GSM-900 [4]	0.90	0.876–0.915	0.921–0.960	955–1023, 0–124
T-GSM-900 [4]	0.90	0.8704–0.876	0.9154–0.921	Dynamic
DCS-1800 [4]	1.80	1.7102–1.7848	1.8052–1.8798	512–885
PCS-1900 [4]	1.90	1.8502–1.9098	1.9302–1.9898	512–810
UMTS [5]		1.885–2.025 1.710–1.755 (US)	2.110–2.200 2.110–2.155 (US)	
802.11 b/g/n (Wi-Fi; ISM) [6]		2.4–2.4835 (ISM)		
802.11 a/h/j (Wi-Fi; UNII) [6]		5.15–5.35 (UNII) 5.47–5.725 5.725–5.825 (ISM/UNII) 4.9–5 GHz, 5.03–5.091 (Japan)		
802.15.4 (Zigbee) [7]		0.898, 0.915 2.4 (ISM)		
802.15.1.1a (Bluetooth) [8]		2.4–2.4835 (ISM)		
802.15.3 (UWB) [9]		Typically > 0.500 GHz bands within the 3.1–10.6 GHz spectrum		
802.16 (Wi-Max) [10]		Various bands within the 2.3–5.8 GHz spectrum		
LTE [11]		Various bands within the 0.45–3.5 GHz spectrum		1–44
5G [12]		Various bands within the 0.66–5.0 GHz spectrum; 26–29 GHz and 37–40 GHz bands		
Internet of Things (IoT) [13]		Various bands within the 0.45–2.2 GHz spectrum		LTE-M and NB-IoT standards

that supports several wireless communications technologies is considered a multiband antenna. For example, an antenna that instantaneously operates two separate bands incorporating frequencies of 380–500 MHz and 1,710–1,990 MHz is referred to as a pentaband antenna because it covers five GSM frequency bands (at 380, 450, 480, 1800, and 1900 MHz). While dealing with the challenges of designing a compact antenna, the designer must also incorporate multiple frequency bands.

This chapter discusses many of the advanced techniques used in the design of multiband antennas. It provides multiband design insight, isolation improvement techniques, applications, and insight into the physical structure of multiband antennas using characteristic mode analysis (CMA).

5.2 TYPES OF MULTIBAND MIMO ANTENNA DESIGNS

The main emphasis of this section is the design techniques used for multiband antennas. The discussion focuses on specific design approaches for achieving multiband MIMO antenna characteristics.

5.2.1 Higher-order resonances

Monopole antennas utilize a $\lambda/4$ length structure to resonate if the electrical field at the input is a minimum and the current is a maximum. An analogous circumstance occurs at the feed when the element size is $3\lambda/4$. Thus, for a fixed antenna dimension, the feed circumstances will be equivalent when the frequency is comparable to heights of $\lambda/4$ and $3\lambda/4$: the monopole will be resonant at an initial frequency and at a frequency three times higher than the initial frequency, as shown in Figure 5.1. Additional sophisticated resonances exist at higher frequencies. This technique is used in numerous types of resonant antennas such as dipoles, patches, slots, dielectric resonators, and monopoles and is repeatedly utilized (and controlled) to achieve multiband behavior [14].

5.2.2 Resonant trap antenna

The traps are either parallel inductor-capacitor circuits or short-circuited transmission line stubs that are designed to be antiresonant (having essentially an infinite input impedance) at some particular frequency. They empower the antenna to work at two different resonant frequencies. At the higher frequency, the trap is adjusted to be close to antiresonance. At

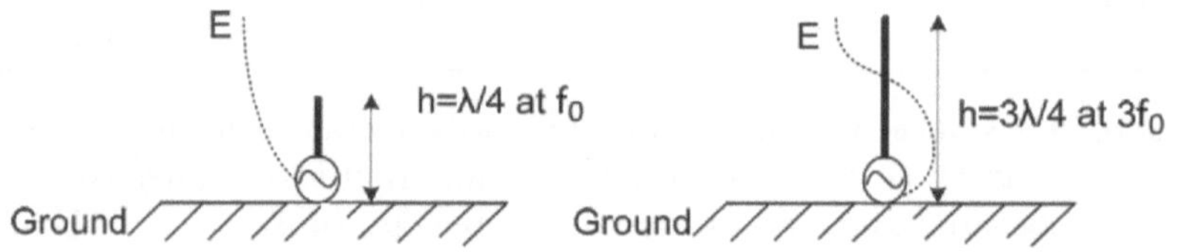

Figure 5.1 Higher-order resonance.

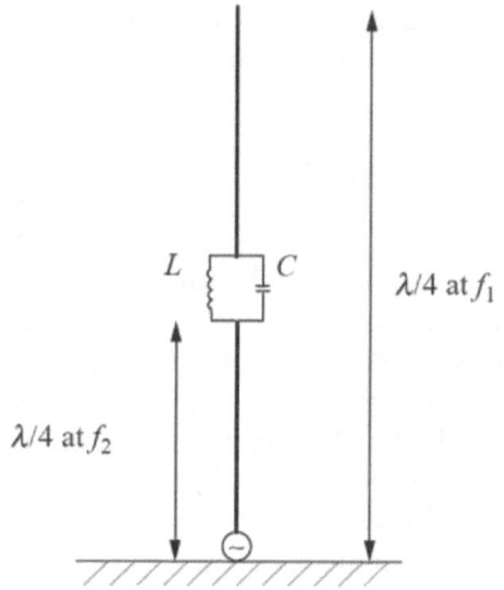

Figure 5.2 Antenna with a resonant trap.

the lower frequency, the trap is well below resonance and is inductive, as shown in Figure 5.2 [15].

5.2.3 Multiple resonant structures

Another common method for achieving multiband operation from an antenna system is to implement numerous resonant configurations, thus utilizing more than one closely located resonant construction with a common feed. As illustrated in Figure 5.3(a), the antenna has two resonant frequencies, f_1 and f_2, for dual-band applications. In this example, a common feed excites two resonant structures simultaneously. Occasionally, various resonant structures can be excited successively, so that the second resonant structure is excited by the first resonant structure, as shown in Figure 5.3(b).

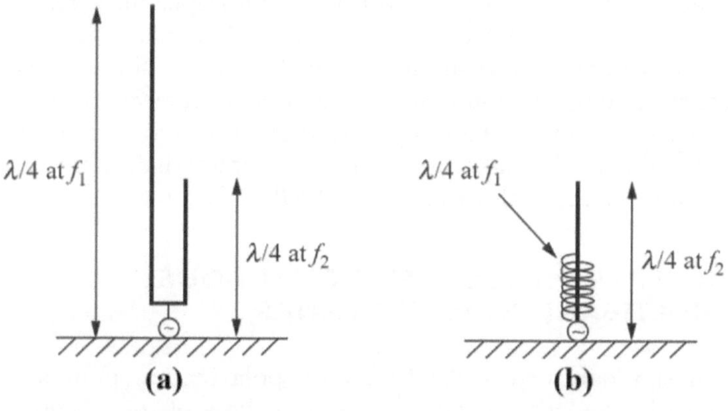

Figure 5.3 Multiple resonant structures.

For example, in [16], the arrangement of monopole and helical antennas provides the dual-band properties of an antenna system designed to obtain GSM 900 and 1800 bands.

5.2.4 Parasitic resonators

In this method, the parasitic element is utilized to achieve multiband characteristics of the antenna structure. An additional auxiliary component is fed by the main element to achieve another operating band. This parasitic element is coupled from the near field of the main antenna, which provides an additional frequency. For example, in [17], a planar inverted-F antenna without a coupling feed has its first resonance at the GSM frequency band and its second resonance at the DCS/PCS frequency band. The s-parameter results of the antenna with and without parasitic elements distinguish the coupling line and parasitic element effect.

5.2.5 Inverted-F antennas

The other popular design is the inverted-F antenna (IFA). It comprises a monopole antenna above and parallel to a ground plane, which is grounded at one end. The intermediate arm of this structure is fed and spaced apart from the grounded end. The design has the advantages of compactness and controlled impedance matching over a simple monopole. An example of an IFA, in which a shorting pin is used to ground the arm and a 50 Ω feed line is used to excite the antenna, appears in [18].

5.2.6 Planar inverted-F antennas

The PIFA can be considered a kind of linear IFA with the wire radiator element replaced by a plate to expand the bandwidth. It is increasingly used in the mobile phone market. The PIFA is resonant at a quarter-wavelength due to the shorting pin at the end. The feed is placed between the open and shorted ends, and the position controls the input impedance. This antenna resembles an inverted F, which explains the PIFA name. It has turned out to be the most popular antenna for handheld devices, since it has several benefits such as compact design, small size, lightweight structure, low manufacturing cost, ability to be integrated with other handset components, good electrical performance, good radiation characteristics, high gain, and low value of the specific absorption rate (SAR) [19].

5.3 ISOLATION IMPROVEMENT TECHNIQUES IN MULTIBAND MIMO ANTENNAS

Antenna design parameters like directivity, polarization, gain, and radiation pattern have identical significance regardless of the antenna design used, so the designer can create a structure to accomplish objectives related to these parameters. For example, a high-directivity antenna can be created as a distinct high-directivity antenna element—for instance, a dish antenna—or perhaps be organized by developing an array with a large number of closely spaced dipoles or patch antennas. Even though the radiation features of the two different designs may be designed to correspond, the antenna design would be required to have multiple input ports. This major change adds flexibility as well as complexity to the design. Moreover, a MIMO antenna radiation pattern could be controlled by tuning the phase and amplitude features of the input signals to the antenna ports. For example, the dual-element dipole MIMO antenna shown in Figure 5.4 could focus the transmitted signal power by changing the phase difference between the two input signals. However, the MIMO antenna design offers an additional challenge to antenna engineers: input power applied to one port of the MIMO elements affects the other elements. This might drastically affect antenna performance because a fraction of the input power could leak toward the radio-frequency circuits through the antenna ports or spread to the other antenna elements. This would cause the performance parameters of the antenna like radiating efficiency and matching to decrease, a result termed mutual coupling. This impact turns out to be more clear as antenna elements are situated nearby. Numerous scientists who design MIMO antennas are working to eliminate mutual coupling issues by using several methods to isolate the antenna elements. This subsection focuses on isolation issues specific to MIMO antenna design approaches used to achieve multiband characteristics. The discussion of these approaches includes examples of types of multiband antennas.

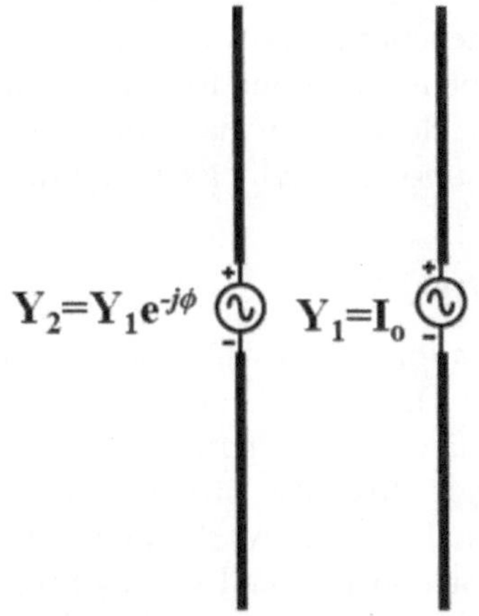

Figure 5.4 Two-element dipole MIMO antenna.

5.3.1 Diversity techniques

5.3.1.1 Spatial diversity

Wireless communication technology's major performance limitation is due to multipath fading. This problem comes into the picture mainly from the destructive addition of multiple paths in the propagation environment. Diversity techniques try to alleviate multipath fading by transmitting multiple replicas of the same information signal independently. This method with multiple independent paths reduces the fading, and higher spectral efficiency (measured by total number of bits per second per Hertz) is reached. Then some arrangement has to be made at the receiver end. In the case of the best path, the idea of spatial diversity applies. Otherwise, in combining all signals to achieve a higher receiver signal-to-noise ratio (SNR), MIMO capacity increases concerning a single antenna [20]. In case of the interference correlation between any two elements, adaptive beamforming or optimum combining can also be used. In MIMO antenna systems, interference can be alleviated naturally with the spatial distance that separates each antenna in the array. The foremost method employed initially to increase capacity was to use an array of spatially separated antennas. Thus, the term *spatial diversity* was invented. Subsequently, there is a certain separating distance in spatial diversity; still, mutual coupling exists between elements and will probably slow the projected growth in capacity.

An example comprises two distinct multiband elements. The first element is a monopole antenna and operates in the 805–1150 and 1780–2170 MHz frequency bands. A PIFA-shaped antenna is the second element, and it operates in the 830–950, 1770–1950, and 2070–2170 MHz frequency bands. In this example, the performance of the compact multiband antenna is improved by utilizing the spatial diversity technique. The two-element antenna separation is 84 mm, working in the WCDMA850, WCDMA1800, and 3G EU (UMTS) bands. The length of the MIMO structure is 100 mm, whereas the ground plane length is only 84 mm. Spatial diversity is a widely used isolation improvement technique for MIMO antennas because of its ease of implementation. However, it has limited design use wherever the separation between monopoles usually goes beyond half a wavelength [21].

5.3.1.2 Pattern diversity

Spatial diversity requires lots of space, which is not always available. To decouple the MIMO system where adequate space is not available, the MIMO platform can make use of pattern diversity [22]. Pattern diversity, also known as angle diversity, provides the required diversity gain by using different beams. Several works define pattern diversity as the natural generalization of conventional orthogonal polarization diversity, since the orthogonality between patterns decorrelates the signals [23]. In some works, pattern diversity is found to result in better performance than spatial diversity when used in handsets [24]. However, this may be limited to situations with relatively large angular spread, such as dense urban environments, making its use in rural scenarios somewhat limited.

A multiband MIMO antenna is presented in [25]. The dual-element MIMO antenna has pattern diversity, which is appropriate for multimode wireless mobile devices with a compact dimension. It functions over an extensive bandwidth, from 890 MHz to 6 GHz, and it contains two monopoles composed of five metal strips, a patch, and a ground plane. The two elements of the MIMO antenna are orthogonally placed for diversity performance.

In another example, a two-port pattern-diversity antenna has been designed for third and fourth generation (3G and 4G) indoor applications [26]. A monocone antenna and a coupled-fed microstrip antenna are used in it. The first element of the antenna, the monocone antenna, is vertically polarized, and the other element, the microstrip antenna, is horizontally polarized.

5.3.1.3 Polarization diversity

In spite of the benefits of pattern diversity and spatial diversity, the possibility of getting extra channels by using orthogonal polarization states has recently gained attention. This is known as polarization diversity. The enhancement allowed by polarization diversity in wireless systems is usually achieved by having a different decorrelated channel delivered by a polarization state orthogonal to the current one. A randomly orientated linearly polarized antenna is also typically used at the receiver to obtain polarization diversity.

A 10-element multimode multiband MIMO antenna for future application in handsets is presented in [27]. The T-shaped slot antenna has a dual-mode property, and it can operate in LTE bands 42, 43, and 46 (3.4–3.8 GHz and 5.15–5.925 GHz). Polarization diversity is achieved by orthogonal placement of the linearly polarized elements.

5.3.2 Isolation improvement techniques

This section presents several multiband MIMO antenna design and isolation improvement methods. All techniques are described individually with examples given so one can better understand the antenna's decoupling in the MIMO system.

5.3.2.1 Defected ground structure

The defected ground structure (DGS) comprises the slots or defects incorporated on the ground plane of antennas or planar circuits [28]. It is a popular method for refining numerous considerations of MIMO antennas, including narrow cross-polarization, bandwidth, tuning frequency, decoupling improvement, outline, multiband operation, and low gain [29, 30]. Additionally, this method considerably improves isolation. However, the back radiation may also be enhanced. The DGS tunes the ground plane surface currents, and may act as a band-stop filter and suppress the coupling between the nearby elements.

A mmWave and microwave band application is presented for a MIMO antenna system in [31]. The MIMO antenna system works in the 2.45, 2.6, 5.2, 24, and 28 GHz bands and has dual-band properties in the 2.45, 2.6, 5.2 GHz bands. In this design, the tapered slot engraved in the ground plane is used to decouple the MIMO elements. Thus, the tapered slot works as a slot antenna at 28 GHz and a decoupling structure at 2.45 GHz [31].

In another example, a multiband MIMO antenna design for 4G and fifth generation (5G) applications is introduced [32]. The structure comprises a dual-element connected antenna array–based MIMO antenna for the 5G band and a dual-element slot-based MIMO antenna system for the 4G band. The ground plane is engraved with two rectangular loops on the boundary. The thin loops act as the 4G MIMO antenna elements, whereas its sides are working as 5G elements.

5.3.2.2 Parasitic element technique

Parasitic elements or etched slit antennas produce dual orthogonal modes by coupling in the ground plane and/or radiating patch, creating a wideband [33]. This technique utilizes supplementary coupling paths to improve isolation among the antenna elements [34, 35]. The dual coupling paths prevent the arriving signal from one element to another, which affects an enhancement in the decoupling level. The decoupling modules, such as the folded shorting strip meandered slot and the vertical parasitic strip, are known as parasitic elements and are obliquely connected [36–38]. The significant advantages of the parasitic or slot antenna are size, easy implementation, and an accessible design process involving the application of waveguides and/or printed circuit board technology. However, the location of the parasitic elements has to be precise, and it is not straightforward.

A compact linearly polarized multiband MIMO antenna system offers the GSM850/900, DCS, PCS, UMTS, WLAN, and WiMAX frequencies. The MIMO antenna consists of two monopoles and a decoupling structure designed using two inverted-L branches and a rectangular slot with one circular end etched on the ground plane, as presented in [39].

5.3.2.3 Neutralization line technique

A lumped component or metallic slot connected from one antenna to another that helps to transfer electromagnetic waves is termed a *neutralization line* [40]. It generates an opposing coupling, which reduces the coupling of particular frequencies among the antenna elements. This approach cancels the coupling by providing an additional path with the same magnitude and inverted phase. These techniques for isolation improvement are suitable for implementation with MIMO antennas having fewer elements.

The dual crossed neutralization lines in a MIMO system using an LTE/WWAN handset are presented in [41]. The antenna system covers the 702–968, 1698–2216, and 2264–3000 MHz bands. It works on all LTE (700, 2300, and 2500), GSM (850 and 900), DCS, PCS, UMTS, and 2.4 GHz Wi-Fi bands. Two crossed neutralization lines with two inductors are hosted to improve isolation between the multiband MIMO antenna elements, particularly in the lower frequencies [41].

In another example, a multiband MIMO antenna is designed using a tri-band monopole antenna in a compact dimension. The antenna can resonate in three bands at the center frequencies of 2.3 GHz, 3.5 GHz, and 5.7 GHz using two branches. To realize high isolation, the U-shaped neutralization line connected with two lines can increase the isolation in a high-frequency band, and the inverted U–shaped neutralization line connected with two patches can decouple in the lower-frequency band [42].

5.3.2.4 Decoupling circuit approach

Decoupling circuits help realize sufficient isolation improvement in MIMO antennas. This effort converts the imaginary part of the mutual admittance of the MIMO antenna system to a matching circuit with the help of the transmission lines. A coupled resonator [43], a manmade structure [44], eigenmode disintegration [45], and introduced components [46] are some methods of the decoupling circuit.

In a dual-band MIMO antenna example, a dual-band antenna is decoupled using a dual-band coupled-resonator decoupling circuit. In this MIMO antenna, the stepped-impedance transformer is used to achieve imaginary mutual admittance in the targeted frequency band. Then a dual-band coupled-resonator decoupling circuit is used to isolate the antenna elements [47]. In another example, a T stub–based decoupling circuit is presented to decouple the dual-band MIMO antenna [48].

5.3.2.5 Other important decoupling techniques

Metamaterial: In this design method, the iterating shape of the materials makes them proficient in influencing electromagnetic waves. Correspondingly, a surface [49] with a negative permeability works as a metamaterial to enhance isolation in the MIMO antenna system. In an example, a dual-band 3-D metamaterial is used to realize low coupling in the MIMO antenna system with dual-band coverage. It comprises two patches, a 3-D metamaterial, and a ground plane. The patch antennas and metamaterial structures are fabricated on different layers, as presented in [50].

Complementary split-ring resonator: These are typically periodic structures shaped like a square, rectangular, or circular ring. They have a strip or gap in the design to carry out filtering as well as to improve isolation among the MIMO antenna elements, as presented in [51].

Electromagnetic bandgap structure: This structure can stop electromagnetic waves of a particular frequency or act as a medium to transmit electromagnetic waves [52]. The periodic structure is made of dielectric or metallic material, and its distinct resonance can produce various bandgaps [52].

5.4 MULTIBAND MIMO ANTENNA DESIGN USING CHARACTERISTIC MODE ANALYSIS

In the previous sections, some multiband MIMO antenna examples were discussed for different applications. The various design methods presented indicate the lack of a suitable design process. A methodical approach for designing an antenna requires physical insight into the structure that the antenna designer can use to achieve the preferred results. The use of CMA has become more widespread among antenna designers, thanks to its ability to determine the physical implications of the structure [53]. CMA helps provide the currents and scattered fields of perfectly conducting bodies regardless of shape and size without any source. These currents and fields are signified using the diverse characteristic modes (CMs) of CMA [54].

5.4.1 Theory of characteristics mode

The CMA theory was developed in 1965 by Garbacz [54]. CMA depends on the structure of the perfectly electrical conductor (PEC) element. According to CMA, the PEC structure radiation pattern is a linear composition of CM patterns. The generalized impedance matrix of the PEC was first given by Harrington and Mautz in 1971 [55, 56]. Furthermore, they recommend determining CMs using the computational technique. The diversified orthogonal set of currents and scattered fields of the subjective PEC structure is called the CM.

$$J = \sum_n a_n J_n \tag{5.1}$$

$$E = \sum_n a_n E_n \tag{5.2}$$

where a_n is a modal weighted coefficient (MWC) and J_n and E_n are characteristic currents (CC) and characteristic fields (CF), respectively.

The current J_n is the nth CM equation given by Equation (5.3) [53, 54].

$$[X][J_n] = \lambda_n [R][J_n] \tag{5.3}$$

where [X] and [R] are the imaginary and real Hermitian parts of a generalized impedance matrix called the reactance and resistance matrix and λ_n is an eigenvalue. The antenna's overall energy is associated with the eigenvalue that is derived from the generalized impedance matrix and extends from −□ to □, which signifies the magnetic energy stored, is beyond 0. If it is below 0, the electrical energy present in the mode and eigenvalues equal to 0 correspond to the resonance condition of the mode. Similarly, characteristic fields E_n can be expanded in the form of the orthogonal equation using the eigenvalue. The modal significance (MS) stabilizes the limit of the eigenvalue in the range from 0 to 1. The resonant frequency of a particular mode, if attained, is $MS = 1$, and the frequency band of that mode emerges at $MS = 0.707$ [57].

$$MS = \left| \frac{1}{1 + j\lambda_n} \right| \tag{5.4}$$

5.4.2 Multiband MIMO antenna design using CMA

Designing a multiband MIMO antenna design with high isolation and impedance bandwidth is a major challenge for today's designers. In [58], the authors present a multiband MIMO antenna that is excited with the orthogonal mode by placing the feed point at the maximum modal current distribution position in the structure. A chassis-mode four-port MIMO antenna design for multiband application is presented in [59]. The inductive and capacitive coupling elements are used to excite the pair of ports to get the desired modes of excitation and high isolation.

5.4.3 Physical understanding of structure

In the example presented in [60], a dual-band MIMO antenna is systematically designed using CMA. The design process of the improved antenna structure is accomplished scientifically using CMA. The modal significance of CMA is used to design the antenna; six modes are examined to acquire physical insight into the structure. From the modal significance results, the designer identifies the resonating and nonresonating modes. The occurrence of unwanted modes increases the mutual coupling near the resonating frequency, requiring modification by the designer.

5.5 MULTIBAND MIMO ANTENNA DESIGN TECHNIQUES FOR DIFFERENT APPLICATIONS

The multiband antenna has been an attractive design for numerous mobile communication devices. Many design techniques have since been developed, as described in the earlier sections. In the following sections, some of the most popular multiband antenna applications are covered.

5.5.1 Vehicular applications

In recent years, interest in wireless technology has increased significantly in the automotive industry. Multiple antennas may be required to cover several important frequency bands for numerous wireless communication applications in future vehicles. In general, advanced vehicles will include LTE, vehicle-to-infrastructure (V2X), and vehicle-to-vehicle technologies to deliver efficient data rates for numerous uses such as voice calls, video streaming, and internet browsing. V2X drives vehicles to communicate with each other to encourage traffic safety and transportation competence in the current infrastructure, which is presently established using IEEE 802.11p. The European Telecommunications Standard Institute (ETSI) uses IEEE 802.11p as the basis for the ITS-G5 standard with a bandwidth allocation of 30 MHz at 5.9 GHz (i.e., from 5875 to 5905 MHz) [61]. The increasing antenna requirement and the restricted space are the result of integrating multiband antennas into a single compact device.

For example, a wideband monopole antenna with an additional resonator is used to achieve all LTE frequency bands on a single radiator. The antenna is attached to a cylindrical substrate that is fabricated using fused filament fabrication (FFF) technology. Low-cost polylactic acid plastic filament (PLA) material is used the antenna in [62]. An eight-element antenna with horizontal and vertical placement is presented in [63]. It incorporates a large number of standards like Bluetooth, WiMAX, WLAN, and ultra-wideband (UWB). The design of the frequency-agile antenna, which reconfigures multibands and the entire UWB frequency band, is presented in [63].

5.5.2 Mobile handsets

In recent years, the high-data-rate requirement has increased exponentially, making significant demands on bandwidth and interoperability within the technologies and requiring wireless devices that can function in different multiband environments. Therefore, the design of a multiband antenna for devices with small dimensions offers a big challenge.

A metal-frame-based multiband four-element MIMO antenna system for a handset for current technology (4G), upcoming technology (5G), and GPS is presented in [64]. The proposed four-element MIMO antenna consists of a primary antenna and a secondary antenna. The primary antenna can cover the LTE (700, 2300, and 2500), GSM (850, 900, 1800, and 1900), and 3.5 GHz frequency bands. The secondary antenna has dual MIMO elements that can operate the GPS, GSM (1800 and 1900), LTE (2300 and 2500), and 3.5 GHz operational bands. Thus, the dual-antenna elements can operate the GPS band and the 698–960 MHz band, whereas the four antenna elements can operate the 1710–2690 and 3400–3600 MHz frequency bands.

Overall, the four-element MIMO antenna system can operate the GPS band; the 2G, 3G, and 4G communication bands; and the 5G band at 3.5 GHz without using any isolating structure between the antenna elements [64].

An example of a compact dual-element MIMO antenna working on DCS, GSM, LTE, PCS, prospective sub-6 GHz 5G applications, and WLAN is given in [65]. The radiating element has a PIFA structure with a quarter-oval shape and defected ground. The antenna displays multiple resonate frequencies: 0.9, 1.8, 1.9, 2.1, 2.3, 2.4, 3.6, and 5.6 GHz (for GSM, DCS/LTE, PCS/UMTS, UMTS, LTE, Wi-Fi, and WLAN).

5.5.3 Laptop/tablet applications

A compact frequency-reconfigurable MIMO antenna ($5 \times 125 \times 1$ mm^3) for laptop applications is presented in [66]. It uses a T-shaped DC line and two PIN diodes (D1 and D2) in conjunction with the proximity-coupled feed structure. Furthermore, the proposed MIMO antenna is reconfigurable, covering the LTE 13 and 17 bands (0.704–0.787 GHz) and the LTE 7 and 20 bands (0.791–0.862 and 2.5–2.69 GHz) using a PIN diode.

5.5.4 IoT applications

An example of a compact planar five-element multi-band MIMO antenna system is presented in [67]. A tunable dual-element folded meandered MIMO antenna operates the LTE frequency bands lower than 1 GHz (687–813 MHz) and the RFID frequencies (2.4 and 5.8 GHz). The other dual-element compact MIMO antenna covers four frequency bands (0.754–0.971, 1.65–1.83, 2–3.6, and 5.1–5.6 GHz). The elements are integrated with a broadband sensing antenna, which senses the spectrum in two frequency bands (0.668–1.94 and 3–4.6 GHz) and works as the ground for the antenna.

A tri-band MIMO antenna, presented in another example, operates the 1.881–1.943, 2.371–2.51, and 3–11 GHz frequency ranges. It offers compatibility when working in multiple bands, including UWB and NB-IoT technologies [68].

5.5.5 5G applications

A dual-band eight-antenna MIMO antenna system for 5G mobile terminals is presented in [69]. It consists of eight L-shaped slot antennas based on stepped impedance resonators (SIRs). The dual resonance can be achieved by changing the impedance ratio of the SIR. The antenna covers LTE band 42 (3400–3600 MHz) and LTE band 46 (5150–5925 MHz).

Another example of a 5G application antenna, a quad-element dual-band MIMO, involves a split-ring resonator (SRR) with inverted-L-shaped antenna elements. The proposed antenna operates around 2.93 GHz, including the lower WLAN and sub-6 GHz 5G bands, and at 5.68 GHz, including the upper WLAN band [70].

In [71], a MIMO antenna with eight elements is presented. This antenna operates in the 3.4–3.6 GHz frequency range. Therefore, its multiband property can cover the sub-6 GHz and mmWave frequency bands. The antenna's square slots are used as a shared radiating structure for both IoT and 5G applications, covering sub-6 GHz and mmWave frequency bands for dual application.

5.6 SUMMARY

In this chapter, a framework of the main challenges and prospects for the design of multiband MIMO antennas is presented. The emphasis is on design techniques, where the most critical issue is the compactness of handheld/portable devices. Strictly speaking, the issue of multiband MIMO antenna systems design can increase with the number of antenna elements. Still, the potential upsurge in antennas is seriously constrained by the small devices of the present era and isolation properties. As a result, several techniques for improving the coupling issues are presented.

Furthermore, to acquire a stronger understanding of the design structure, its physical aspects are explained using CMA in the context of multiband design. Finally, several multiband MIMO antenna systems, using numerous methods and designed for various applications, are considered. The importance of handheld/portable devices has encouraged innovative growth, and therefore many thrilling challenges remain to be addressed in the upcoming generation of technology.

REFERENCES

[1] T. Zahariadis, K. Vaxevanakis, C. Tsantilas, N. Zervos, and N. Nikolaou, "Global Roaming in Next-Generation Networks," *IEEE Commun. Mag.*, vol. 40, pp. 145–151, 2002.

[2] H. Okazaki, A. Fukuda, K. Kawai, T. Furuta, and S. Narahashi, "MEMS-Based Reconfigurable RF Front-End Architecture for Future Band-Free Mobile Terminals," *2007 European Microwave Conference*, Munich, pp. 1058–1061, 2007.

[3] I. Nam, H. Moon, J.-D. Bae, and B.-H. Park, "A Wideband CMOS RF Front-End Using AC-Coupled Current Mirrored Technique for Multiband Multistandard Mobile TV Tuners," *IEEE Microw. Wirel. Compon. Lett.*, vol. 17, pp. 739–741, 2007.

[4] https://rfmw.em.keysight.com/wireless/helpfiles/n7602b/Content/Main/Frequency_Bands_and_ARFCN.htm
[5] https://www.electronics-notes.com/articles/connectivity/3g-umts/frequency-bands-channels-uarfcn.php
[6] https://en.wikipedia.org/wiki/IEEE_802.11#:~:text=IEEE%20802.11%20is%20part%20of,2.4%20GHz%2C%205%20GHz%2C%206
[7] M. B. Mantri, P. Velagapudi, B. C. Eravatri, and V. V. Mani, "Performance Analysis of 2.4GHz IEEE 802.15.4 PHY Under Various Fading Channels," *2013 International Conference on Emerging Trends in Communication, Control, Signal Processing and Computing Applications*, Bangalore, pp. 1–4, 2013, doi: 10.1109/C2SPCA.2013.6749420
[8] "IEEE Standard for Telecommunications and Information Exchange between Systems—LAN/MAN—Specific Requirements—Part 15: Wireless Medium Access Control (MAC) and Physical Layer (PHY) Specifications for Wireless Personal Area Networks (WPANs)," in *IEEE Std. 802.15.1-2002*, pp.1–473, 14 June 2002, doi: 10.1109/IEEESTD.2002.93621
[9] https://en.wikipedia.org/wiki/Ultra-wideband
[10] https://www.electronics-notes.com/articles/connectivity/wimax/frequencies-spectrum-bands.php#:~:text=The%20IEEE%20802.16%20WiMAX%20standard,range%202%20to%2011%20GHz
[11] https://en.wikipedia.org/wiki/LTE_frequency_bands
[12] https://www.cablefree.net/wirelesstechnology/4glte/5g-frequency-bands-lte/
[13] https://www.cablefree.net/wirelesstechnology/internet-of-things-iot./
[14] D. Wu, S. W. Cheung, and T. I. Yuk, "A Compact and Low-Profile Loop Antenna with Multiband Operation for Ultra-Thin Smartphones," *IEEE Trans. Antennas Propag.*, vol. 63, no. 6, pp. 2745–2750, June 2015, doi:10.1109/TAP.2015.2412962
[15] W. O. Coburn and C. Fazi, "A Lumped-Circuit Model for a Triband Trapped Dipole Array—Part 1: Model Description," *IEEE Antennas Wirel. Propag. Lett.*, vol. 8, pp. 14–18, 2009, doi: 10.1109/LAWP.2008.917604
[16] P. Haapala, P. Vainikainen, and P. Eratuuli, "Dual Frequency Helical Antennas for Handsets," *Proceedings of Vehicular Technology Conference*, Atlanta, GA, vol. 1, pp. 336–338, 1996, doi: 10.1109/VETEC.1996.503464
[17] K. Kim, S. Lee, B. Kim, J. Jung, and Y. J. Yoon, "Small Antenna with a Coupling Feed and Parasitic Elements for Multiband Mobile Applications," *IEEE Antennas Wirel. Propag. Lett.*, vol. 10, pp. 290–293, 2011, doi: 10.1109/LAWP.2011.2139190
[18] Y.-L. Kuo, T.-W. Chiou, and K. L. Wong, "A Novel Dual-Band Printed Inverted-F Antenna," *Microw. Opt. Technol. Lett.*, vol. 31, no. 5, pp. 353–355, Dec. 2001.
[19] M. El Halaoui, A. Kaabal, H. Asselman, S. Ahyoud, and A. Asselman, "Multiband Planar Inverted-F Antenna with Independent Operating Bands Control for Mobile Handset Applications," *Int. J. Antennas Propag.*, vol. 2017, Article ID 8794039, 2017, https://doi.org/10.1155/2017/8794039
[20] Z. Chen, et al., "Analysis of Transmit Antenna Selection/Maximal-Ratio Combining in Rayleigh Fading Channels," *IEEE Trans. Veh. Technol.*, vol. 45, no. 4, pp. 1312–1321, July 2005.
[21] J. Avendal, Z. Ying, and B. K. Lau, "Multiband Diversity Antenna Performance Study for Mobile Phones," *2007 International Workshop on Antenna Technology: Small Antennas and Novel Metamaterials*, Cambridge, pp. 193–196, 2007, doi: 10.1109/IWAT.2007.370109

[22] R. G. Vaughan, "Switched Parasitic Elements for Antenna Diversity," *IEEE Trans. Antennas Propag.*, vol. 47, no. 2, pp. 399–405, Feb. 1999.
[23] L. Dong et al., "Multiple-Input Multiple-Output Wireless Communication Systems Using Antenna Pattern Diversity," Global Telecommunications Conference, Taipei, vol. 1, pp. 997–1001, 2002.
[24] P. Mattheijssen et al., "Antenna-Pattern Diversity versus Space Diversity for Use at Handhelds," *IEEE Trans. Veh. Technol.*, vol. 53, no. 4, pp. 1035–1042, July 2004.
[25] Q. Rao and K. Wilson, "Design, Modeling, and Evaluation of a Multiband MIMO/Diversity Antenna System for Small Wireless Mobile Terminals," *IEEE Trans. Compon. Packag. Manuf. Technol.*, vol. 1, no. 3, pp. 410–419, Mar. 2011.
[26] L. Sun, W. Huang, B. Sun, Q. Sun, and J. Fan, "Two-Port Pattern Diversity Antenna for 3G and 4G MIMO Indoor Applications," *IEEE Antennas Wirel. Propag. Lett.*, vol. 13, pp. 1573–1576, 2014, doi:10.1109/LAWP.2014.2346393
[27] Y. Li, C. Sim, Y. Luo, and G. Yang, "Multiband 10-Antenna Array for Sub-6 GHz MIMO Applications in 5-G Smartphones," *IEEE Access*, vol. 6, pp. 28041–28053, 2018, doi:10.1109/ACCESS.2018.2838337
[28] M. K. Khandelwal, B. K. Kanaujia, and S. Kumar, "Defected Ground Structure: Fundamentals, Analysis, and Applications in Modern Wireless Trends," *Int. J. Antennas Propag.*, vol. 2017, Feb. 2017, Article ID 2018527.
[29] A. K. Arya, M. V. Kartikeyan, and A. Patnaik, "Defected Ground Structure in the Perspective of Microstrip Antennas: A Review," *Frequenz*, vol. 64, no. 5–6, pp. 79–84, June 2010.
[30] P. R. Prajapati, "Application of Defected Ground Structure to Suppress Out-of-Band Harmonics for WLAN Microstrip Antenna," *Int. J. Microw. Sci. Technol.*, vol. 2015, Nov. 2015, Article ID 210608.
[31] M. Ikram, N. Nguyen-Trong, and A. Abbosh, "Multiband MIMO Microwave and Millimeter Antenna System Employing Dual-Function Tapered Slot Structure," *IEEE Trans. Antennas Propag.*, vol. 67, no. 8, pp. 5705–5710, Aug. 2019, doi: 10.1109/TAP.2019.2922547
[32] M. S. Sharawi, M. Ikram, and A. Shamim, "A Two Concentric Slot Loop Based Connected Array MIMO Antenna System for 4G/5G Terminals," *IEEE Trans. Antennas Propag.*, vol. 65, no. 12, pp. 6679–6686, Dec. 2017, doi: 10.1109/TAP.2017.2671028
[33] C.-X. Mao and Q.-X. Chu, "Compact Coradiator UWB-MIMO Antenna with Dual Polarization," *IEEE Trans. Antennas Propag.*, vol. 62, no. 9, pp. 4474–4480, Sep. 2014.
[34] S. T. Fan, Y. Z. Yin, B. Lee, W. Hu, and X. Yang, "Bandwidth Enhancement of a Printed Slot Antenna with a Pair of Parasitic Patches," *IEEE Antennas Wirel. Propag. Lett.*, vol. 11, pp. 1230–1233, 2012.
[35] J.-S. Row and S.-W. Wu, "Circularly-Polarized Wide Slot Antenna Loaded with a Parasitic Patch," *IEEE Trans. Antennas Propag.*, vol. 56, no. 9, pp. 2826–2832, Sep. 2008.
[36] Z. Li, Z. Du, M. Takahashi, K. Saito, and K. Ito, "Reducing Mutual Coupling of MIMO Antennas with Parasitic Elements for Mobile Terminals," *IEEE Trans. Antennas Propag.*, vol. 60, no. 2, pp. 473–481, Feb. 2012.

[37] M. Ayatollahi, Q. Rao, and D. Wang, "A Compact, High Isolation and Wide Bandwidth Antenna Array for Long Term Evolution Wireless Devices," *IEEE Trans. Antennas Propag.*, vol. 60, no. 10, pp. 4960–4963, Oct. 2012.
[38] S. Soltani and R. D. Murch, "A Compact Planar Printed MIMO Antenna Design," *IEEE Trans. Antennas Propag.*, vol. 63, no. 3, pp. 1140–1149, Mar. 2015.
[39] S. Shoaib, I. Shoaib, N. Shoaib, X. Chen, and C. G. Parini, "Design and Performance Study of a Dual-Element Multiband Printed Monopole Antenna Array for MIMO Terminals," *IEEE Antennas Wirel. Propag. Lett.*, vol. 13, pp. 329–332, 2014, doi: 10.1109/LAWP.2014.2305798
[40] Y. Wang and Z. Du, "A Wideband Printed Dual-Antenna with Three Neutralization Lines for Mobile Terminals," *IEEE Trans. Antennas Propag.*, vol. 62, no. 3, pp. 1495–1500, Mar. 2014.
[41] S. Wang and Z. Du, "Decoupled Dual-Antenna System Using Crossed Neutralization Lines for LTE/WWAN Smartphone Applications," *IEEE Antennas Wirel. Propag. Lett.*, vol. 14, pp. 523–526, 2015, doi: 10.1109/LAWP.2014.2371020
[42] R. Liu, X. An, H. Zheng, M. Wang, Z. Gao, and E. Li, "Neutralization Line Decoupling Tri-Band Multiple-Input Multiple-Output Antenna Design," *IEEE Access*, vol. 8, pp. 27018–27026, 2020, doi: 10.1109/ACCESS.2020.2971038
[43] L. Zhao, L. K. Yeung, and K.-L. Wu, "A Novel Second-Order Decoupling Network for Two-Element Compact Antenna Arrays," 2012 Asia Pacific Microwave Conference Proceedings, Kaohsiung, Taiwan, pp. 1172–1174, 2012, doi: 10.1109/APMC.2012.6421860
[44] J. Andersen and H. Rasmussen, "Decoupling and Descattering Networks for Antennas," *IEEE Trans. Antennas Propag.*, vol. 24, no. 6, pp. 841–846, Nov. 1976.
[45] L. K. Yeung and Y. E. Wang, "Mode-Based Beamforming Arrays for Miniaturized Platforms," *IEEE Trans. Microw. Theory Techn.*, vol. 57, no. 1, pp. 45–52, Jan. 2009.
[46] K.-L. Wong, C.-Y. Tsai, and J.-Y. Lu, "Two Asymmetrically Mirrored Gap-Coupled Loop Antennas as a Compact Building Block for Eight-Antenna MIMO Array in the Future Smartphone," *IEEE Trans. Antennas Propag.*, vol. 65, no. 4, pp. 1765–1778, Apr. 2017.
[47] L. Zhao and K. Wu, "A Dual-Band Coupled Resonator Decoupling Network for Two Coupled Antennas," *IEEE Trans. Antennas Propag.*, vol. 63, no. 7, pp. 2843–2850, July 2015, doi: 10.1109/TAP.2015.2421973
[48] J. Sui and K. Wu, "A General T-Stub Circuit for Decoupling of Two Dual-Band Antennas," *IEEE Trans. Microw. Theory Tech.*, vol. 65, no. 6, pp. 2111–2121, June 2017, doi: 10.1109/TMTT.2017.2647951
[49] B. Mohamadzade and M. Afsahi, "Mutual Coupling Reduction and Gain Enhancement in Patch Array Antenna Using a Planar Compact Electromagnetic Bandgap Structure," *IET Microw., Antennas Propag.*, vol. 11, no. 12, pp. 1719–1725, 2017.
[50] S. Luo and Y. Li, "A Dual-Band Antenna Array with Mutual Coupling Reduction Using 3D Metamaterial Structures," *2018 International Symposium on Antennas and Propagation*, Busan, South Korea, 2018, pp. 1–2.

[51] T. Yue, Z. H. Jiang, and D. H. Werner, "A Compact Metasurface-Enabled Dual-Band Dual-Circularly Polarized Antenna Loaded with Complementary Split Ring Resonators," *IEEE Trans. Antennas Propag.*, vol. 67, no. 2, pp. 794–803, Feb. 2019, doi: 10.1109/TAP.2018.2882616

[52] X. Tan, W. Wang, Y. Wu, Y. Liu, and A. A. Kishk, "Enhancing Isolation in Dual-Band Meander-Line Multiple Antenna by Employing Split EBG Structure," *IEEE Trans. Antennas Propag.*, vol. 67, no. 4, pp. 2769–2774, Apr. 2019, doi: 10.1109/TAP.2019.2897489

[53] R. J. Garbacz. "Modal Expansions for Resonance Scattering Phenomena," *Proceedings of the IEEE*, vol. 53, no. 8, pp. 856–864, 1965.

[54] R. Harrington and J. Mautz. "Theory of Characteristic Modes for Conducting Bodies," *IEEE Trans. Antennas Propag.*, vol. 19, no. 5, pp. 622–628, Sep. 1971.

[55] R. Harrington and J. Mautz: "Computation of Characteristic Modes for Conducting Bodies," *IEEE Tran. Antennas Propag.*, vol. 19, no. 5, pp. 629–639, Sep. 1971.

[56] W. Wu and Y. P. Zhang, "Analysis of Ultra-Wideband Printed Planar Quasi-Monopole Antennas Using the Theory of Characteristic Modes," *IEEE Antennas Propag. Mag.*, vol. 52, no. 6, Dec. 2010.

[57] Yikai Chen and Chao-Fu Wang, *Characteristic Modes: Theory and Applications in Antenna Engineering* (Hoboken, NJ, Wiley, 2015).

[58] Z. Miers, H. Li, and B. Lau, "Design of Bandwidth-Enhanced and Multiband MIMO Antennas Using Characteristic Modes," *IEEE Antennas Wirel. Propag. Lett.*, vol. 12, pp. 1696–1699, 2013.

[59] K. Kishor and S. Hum, "Multiport Multiband Chassis-Mode Antenna Design Using Characteristic Modes," *IEEE Antennas Wirel. Propag. Lett.*, vol. 16, pp. 609–612, 2017.

[60] H. V. Singh, S. Tripathi, and R. Vaddi, "Systematic Design of Dual-Band MIMO Antenna Using Characteristic Mode Analysis," *2018 5th IEEE Uttar Pradesh Section International Conference on Electrical, Electronics and Computer Engineering*, Gorakhpur, pp. 1–6, 2018, doi: 10.1109/UPCON.2018.8596915

[61] E. G. Ström, "On Medium Access and Physical Layer Standards for Cooperative Intelligent Transport Systems in Europe," *Proceedings of the IEEE*, vol. 99, no. 7, pp. 1183–1188, July 2011.

[62] B. Sanz-Izquierdo, S. Jun, and T. B. Baydur, "MIMO LTE Vehicular antennas on 3d Printed Cylindrical Forms," *Wideband and Multi-band Antennas and Arrays for Civil, Security and Military Applications*, London, pp. 1–5, 2015, doi: 10.1049/ic.2015.0145

[63] M. G. N. Alsath et al., "An Integrated Tri-Band/UWB Polarization Diversity Antenna for Vehicular Networks," *IEEE Trans. Veh. Technol.*, vol. 67, no. 7, pp. 5613–5620, July 2018, doi: 10.1109/TVT.2018.2806743

[64] D. Huang, Z. Du, and Y. Wang, "A Quad-Antenna System for 4G/5G/GPS Metal Frame Mobile Phones," *IEEE Antennas Wirel. Propag. Lett.*, vol. 18, no. 8, pp. 1586–1590, Aug. 2019, doi: 10.1109/LAWP.2019.2924322

[65] H. V. Singh, S. Tripathi, and R. Vaddi, "A Compact Planner MIMO Antenna for GSM, DCS, PCS, UMTS, LTE, WLAN and 5G Application," *2018 International Conference on Advances in Computing, Communication Control and Networking*, Noida, India, pp. 1068–1073, 2018, doi: 10.1109/ICACCCN.2018.8748510

[66] B. Mun, C. Jung, M. Park, and B. Lee, "A Compact Frequency-Reconfigurable Multiband LTE MIMO Antenna for Laptop Applications," *IEEE Antennas Wirel. Propag. Lett.*, vol. 13, pp. 1389–1392, 2014, doi:10.1109/LAWP.2014.2339802

[67] K. R. Jha, B. Bukhari, C. Singh, G. Mishra, and S. K. Sharma, "Compact Planar Multistandard MIMO Antenna for IoT Applications," *IEEE Trans. Antennas Propag.*, vol. 66, no. 7, pp. 3327–3336, July 2018, doi: 10.1109/TAP.2018.2829533

[68] N. K. Maurya and R. Bhattacharya, "Design of Compact Dual-Polarized Multiband MIMO Antenna Using Near-Field for IoT," *AEU—Int. J. Electron. Commun.*, vol. 117, Article ID 153091, 2020, doi: 10.1016/j.aeue.2020.153091

[69] J. Li et al., "Dual Band Eight-Antenna Array Design for MIMO Applications in 5G Mobile Terminals," *IEEE Access*, vol. 7, pp. 71636–71644, 2019, doi: 10.1109/ACCESS.2019.2908969

[70] D. Sarkar and K. V. Srivastava, "Compact Four-Element SRR-Loaded Dual-Band MIMO Antenna for WLAN/WiMAX/WiFi/4G-LTE and 5G Applications," *Electron. Lett.*, vol. 53, no. 25, pp. 1623–1624, Dec. 2017.

[71] S. H. Kiani, A. Altaf, M. Abdullah, F. Muhammad, N. Shoaib, M. R. Anjum, R., Damaševičius, and T. Blažauskas, "Eight Element Side Edged Framed MIMO Antenna Array for Future 5G Smart Phones," *Micromachines*, vol. 11, p. 956, 2020, doi: 10.3390/mi11110956

Chapter 6

MIMO antenna in UWB applications

Atanu Chowdhury, Tejaswita Kumari, and Prabir Ghosh

CONTENTS

6.1 INTRODUCTION

This chapter presents a brief discussion about the design, progress, and choices of antenna elements to be used in a multiple-input multiple-output (MIMO) microstrip antenna with ultra-wideband (UWB) application. In a MIMO wireless communication system, both the transmitter and the receiver terminals use antenna arrays.

The chapter focuses on the UWB MIMO antenna design. It also briefly presents a design and testing comparison between a simple UWB antenna and a UWB MIMO antenna.

Currently, the frequencies ranging from 3.1 to 10.6 GHz assigned by the U.S. Federal Communications Commission [1], the ultra-wideband frequencies, are in high demand because of their conventional advantages like high-data-rate transmission, low power spectral density, and high-precision ranging [2] and their attractive characteristics. Today it is a popular topic for researchers working on UWB MIMO antennas. After being introduced in

DOI: 10.1201/9781003290230-6

1950, microstrip antennas became very popular in the 1970s. These antennas are made of a conducting material called a patch on a substrate and ground. The conducting patch may have many different shapes. However, rectangular, square, and circular patches are most in demand because of their ease of fabrication and analysis and their radiation patterns. Microstrip antennas are cost effective, weightless, and easy to feed; have low profile, versatile resonance frequency, polarization, and impedance matching; and are simple to fabricate using printed circuit board (PCB) technology.

The microstrip patch antenna is now broadly used in mobile communication, satellite communication, biomedical applications, and radar because of its inherent characteristics. Various types of substrate material having a dielectric constant (ε_r) range of $2.2 \leq \varepsilon_r \leq 12$ can be used [3].

6.1.1 Significance of multiple antennas

The environment of radio frequency over electrically small aperture platforms is changing at an alarming rate. Until recently, only one radio channel was used in the propagation of radiofrequency (RF) waves, and it was usually connected to only a single antenna. But with modern technology, the conditions are very challenging, as more than a single radio channel is used simultaneously. For example, a cellular phone may have three to five different frequency bands for GPS, Bluetooth, Wireless Fidelity (Wi-Fi), etc. At some points Wireless local area network (WLAN) radios are also available. This means that more complex filtering of RF signals is mandatory. It is now becoming very common for each of the radio channels to apply more than a single antenna so that it can create various uses on MIMO applications. The changing situation is described in Figure 6.1.

MIMO antenna diversion technique is already in use with WLAN radio to counter multipath propagation, reduce outages, and improve the quality

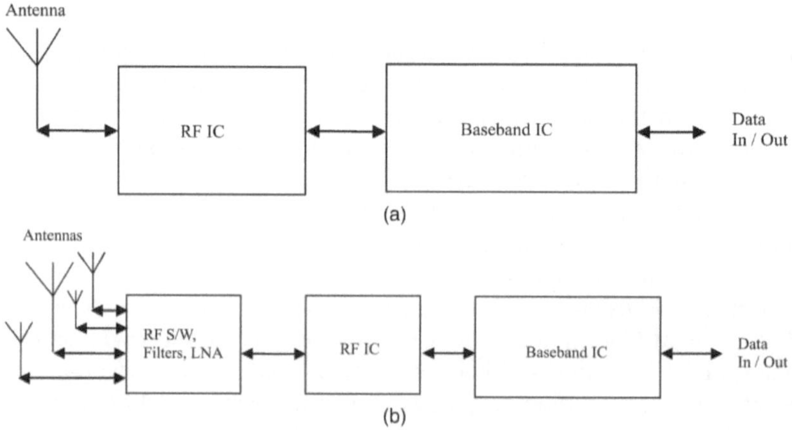

Figure 6.1 Block diagrams of (a) previous and (b) current technology.

and reliability of the RF channel communication links. Three kinds of diversity are apparently used. Two patch antennas may be located as far apart as possible to have some spatial diversity, they should be placed orthogonally to provide polarization diversity, and they may also have diversity in their radiation patterns. Diversity in modern WLAN systems generally comprises two antennas, as it is enough to ensure that if an antenna is at an RF null condition, the other antenna is generally not, which provides a better result in multipath propagations. One radio channel is present, and the receiver antenna receives signals from one antenna at a time. A RF switch is applied to switch to the antenna providing the better signal.

While simple diversity switching tries to minimize the multipath propagation effects, MIMO is used in multipath propagation to improve the quality as well as the reliability of the signal. A MIMO system generally uses more than one antenna—typically, two or four—which is a more efficient technique than diversity switching for the improvement of communication links. This technique is very efficient and powerful and can be further improved by employing a defected ground structure, which is generally used to provide wideband. The UWB MIMO system is more complex and more expensive than the diversity switching system, but it also offers more benefits. The basic formulas for designing a microstrip antenna are given below [3]:

a. Effective dielectric constant

$$\varepsilon_{reff} = \frac{\varepsilon_r + 1}{2} + \frac{\varepsilon_r - 1}{2}\left[1 + 12\frac{h}{w}\right]^{-\frac{1}{2}} \tag{6.1}$$

where h = height of substrate, w = width of microstrip center conductor, and ε_r = relative dielectric constant of dielectric substrate (board material).

b. Width

$$W = \frac{1}{2f_r\sqrt{\mu_0\varepsilon_0}}\sqrt{\frac{2}{\varepsilon_r + 1}} = \frac{\upsilon_0}{2f_r}\sqrt{\frac{2}{\varepsilon_r + 1}} \tag{6.2}$$

where f_r = resonance frequency, υ_0 = speed of light in free space, μ_0 = permeability of free space, and ε_0 = permittivity of free space.

c. Effective length

$$\frac{\Delta L}{h} = 0.412\frac{\left(\varepsilon_{reff} + 0.3\right)\left(\frac{W}{h} + 0.264\right)}{\left(\varepsilon_{reff} - 0.258\right)\left(\frac{W}{h} + 0.8\right)} \tag{6.3}$$

$$L_{eff} = L + 2\Delta L \tag{6.4}$$

where L = length of microstrip center conductor.

d. Resonance frequency

$$(f_r)_{010} = \frac{1}{2L\sqrt{\varepsilon_r}\sqrt{\mu_0\varepsilon_0}} = \frac{\upsilon_0}{2L\sqrt{\varepsilon_r}} \quad (6.5)$$

6.1.2 Basic methodology and techniques of MIMO antenna

For transmitters and receivers, MIMO systems use multiple antennas. They have the capability of combining single-input multiple-output (SIMO) and multiple-input single-output (MISO) technology. The capacity of a MIMO system can also be increased by using spatial multiplexing (SM) [4].

6.1.2.1 SIMO antenna techniques

The system where one antenna transmits signals and multiple antennas receive signals is called SIMO [5]. The SIMO antenna uses a receiver diversity technique where the transmitted signal is received with different strengths at different receivers, as shown in Figure 6.2. The following techniques can be implemented to integrate the received signal:

- Maximum ratio combining (MRC)
- Equal gain combining (EGC)
- Selection combining (SC)

6.1.2.2 MISO antenna techniques

In this antenna technique, multiple antennas are used as transmitters, and a single antenna is used as a receiver [6]. It is considered a modern technology where multiple antennas are needed in the base station (BS). In MISO,

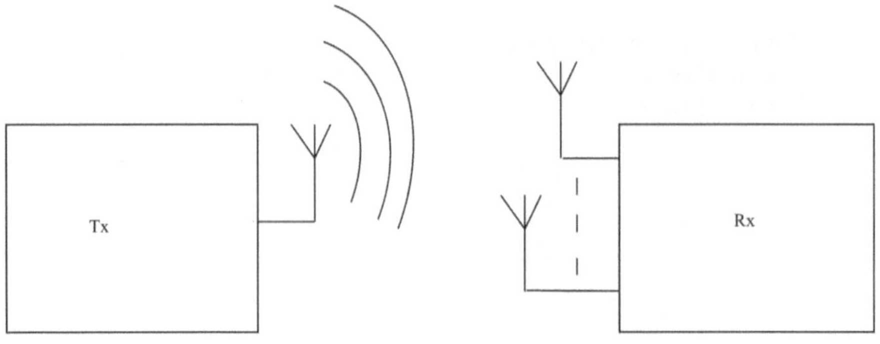

Figure 6.2 Block diagram of SIMO system.

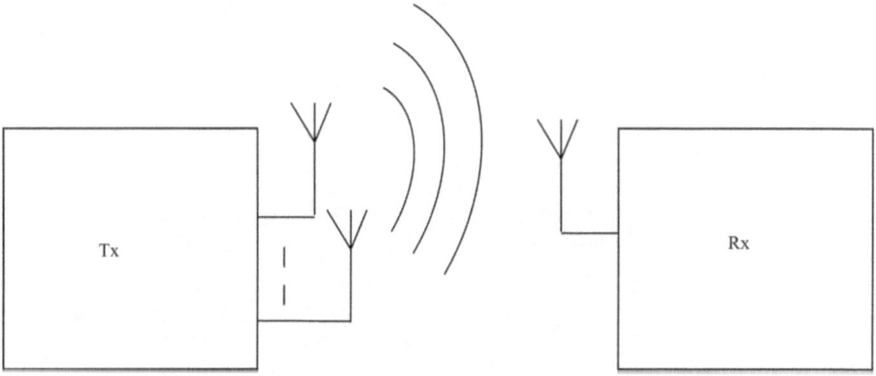

Figure 6.3 Block diagram of MISO system.

the transmit diversity technique is used to produce spatial repetition for transmitted signals via various antennas. A typical MISO system is shown in Figure 6.3.

6.2 ANTENNA DESIGN AND ANALYSIS

In this chapter, the UWB antenna and UWB MIMO antenna are compared and the analysis on previous work [7] of microstrip antenna and few techniques are concentrated [8]. The all UWB antenna can be designed with different shapes and topologies to meet the desired requirements.

6.2.1 Design and analysis of a UWB antenna

The proposed UWB antenna is designed on a low-cost Rogers RT/Duroid 5880 substrate having dimensions of 34 × 30 × 0.813 mm^3, a dielectric constant (ε_r) of 2.2, and a loss tangent (tan δ) of 0.0009. The design of the antenna is very simple [9]. It is basically a rectangular patch connected with two small rectangular studs by a very thin line 0.01 mm wide and 1.5 mm long (see Figure 6.4(a)). The main rectangular antenna is 16.93 × 13.47 mm^2. The two rectangular studs [10] have the same dimensions: 16.93 × 1.00 mm^2. All the rectangular studs, thin line, and main patch have a thickness of 0.035 mm on the substrate. The ground is placed below the substrate, which is defected to provide UWB. It has a thickness of 0.035 mm below the substrate. The defected ground is 34 × 7.56 mm^2 and is placed right below the lumped port connecting the ground and microstrip feed line. The defected ground also consists of a rectangular cavity of 2.2666 × 1.56 mm^2, depicted in Figure 6.4(b). The microstrip feed line has a width of 2 mm, which implies that the characteristic impedance of the line is 42.6 Ω. The feed line is 9 mm long. The lumped port also has an impedance of 42.6 Ω,

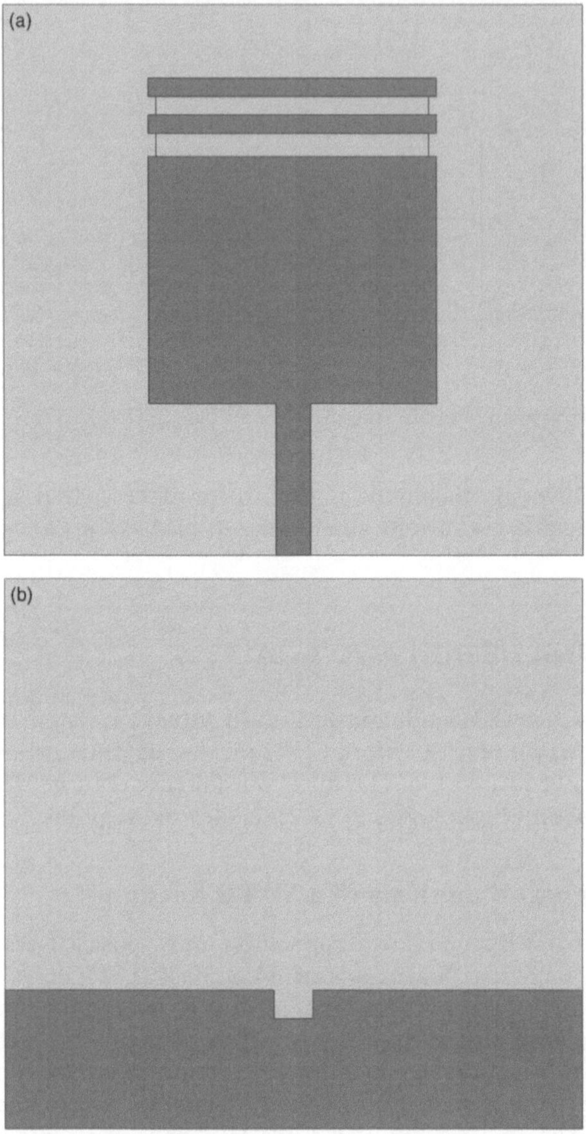

Figure 6.4 (a) Top view and (b) Bottom view of the proposed UWB antenna.

which gives the proposed antenna perfect impedance matching, which improves its performance. Figure 6.5 is a 3-D view of the antenna.

6.2.1.1 Simulated results and analysis

The designed antenna has been simulated by ANSYS Electronics Desktop (High Frequency Structure Simulator [HFSS]) version 21.0 software. The simulated results are shown below.

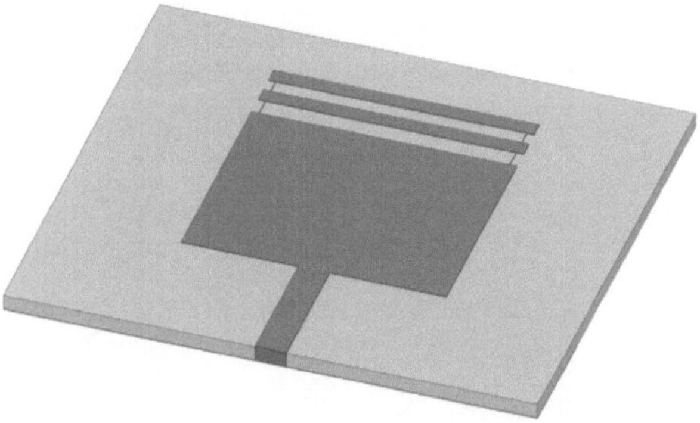

Figure 6.5 3-D view of the proposed UBW antenna.

6.2.1.1.1 S_{11} parameter (return loss)

Figure 6.6 shows that the antenna has two resonating frequencies of 4.6 GHz and 7.1 GHz with return losses of –32.416 dB and –18.426 dB, respectively, which are very acceptable in the field of microwave wireless communication systems. It also provides two wide bands, one of 1.8 GHz from 3.4 to 5.2 GHz, and one of 0.9 GHz from 6.8 to 7.7 GHz. In these bands, the return loss is less than –10 dB in the entire range. It signifies that more than 90 percent of the input power is delivered to the output power, so the power loss is very small.

6.2.1.1.2 Voltage standing wave ratio

Figure 6.7 shows that the two resonating frequencies have voltage standing wave ratios (VSWRs) of 1.049 and 1.272. The frequency bands of 1.8 GHz, ranging from 3.4 to 5.2 GHz, and 0.9 GHz, ranging from 6.8 to 7.7 GHz,

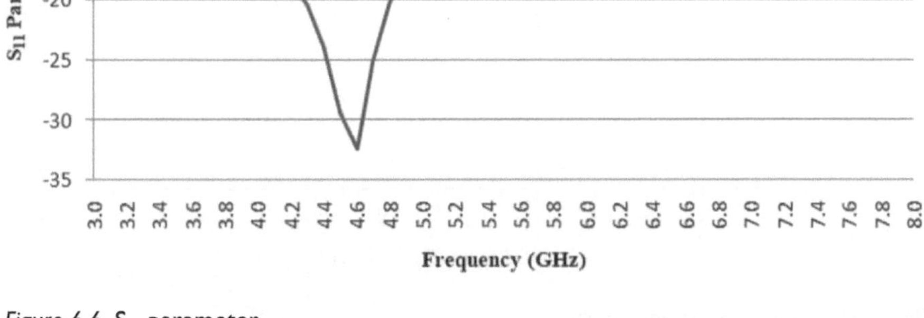

Figure 6.6 S_{11} parameter.

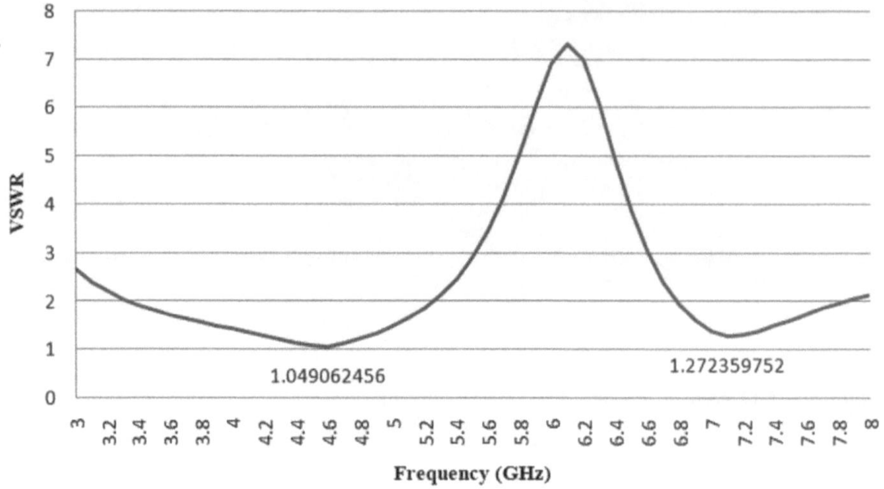

Figure 6.7 VSWR.

also have VSWRs between 1 and 2. This signifies that the impedance matching is well achieved and is very much in the acceptable range.

6.2.1.1.3 Gain vs. frequency

Figure 6.8 shows that the two resonating frequencies have higher gains compared to the other frequencies. The frequency bands of 1.19 GHz, ranging

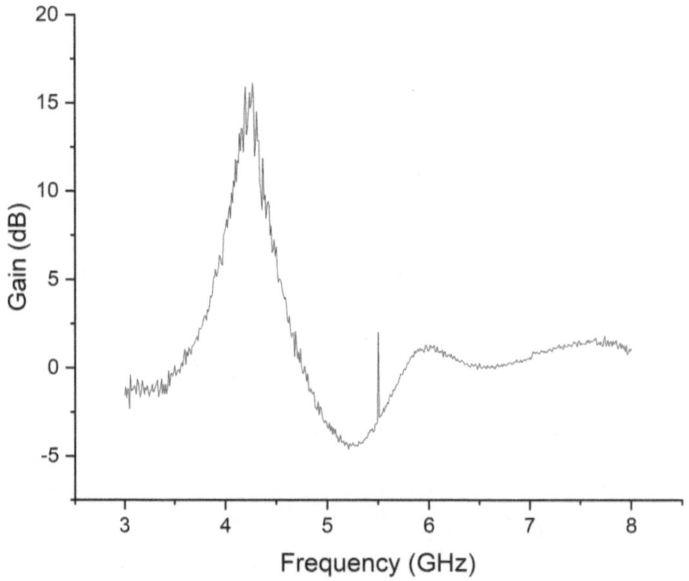

Figure 6.8 Gain vs. frequency curve.

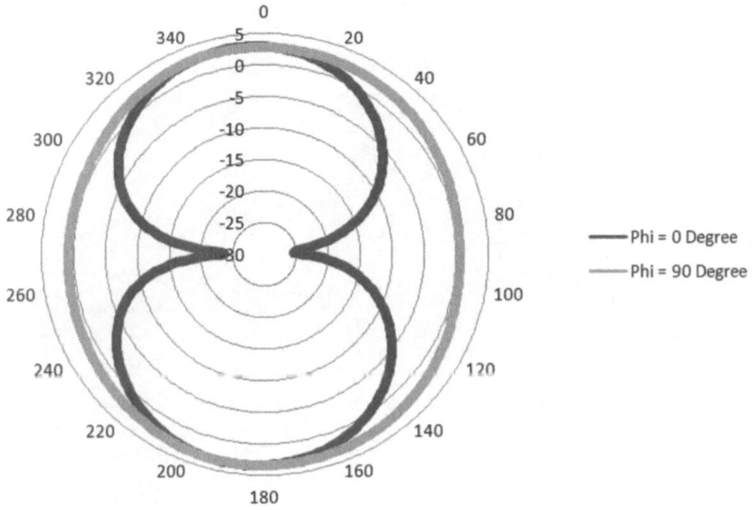

Figure 6.9 Radiation pattern at 4.6 GHz for phi = 0° and 90°.

from 3.56 to 4.75 GHz, and 0.9 GHz, ranging from 6.8 to 7.7 GHz, have positive gains, where other frequencies have negative gains.

6.2.1.1.4 Radiation pattern

The radiation patterns of the proposed UWB antenna at 4.6 and 7.1 GHz are shown in Figures 6.9 and 6.10, respectively. The bidirectional characteristics can be seen for both the frequencies when the angle phi is 0°.

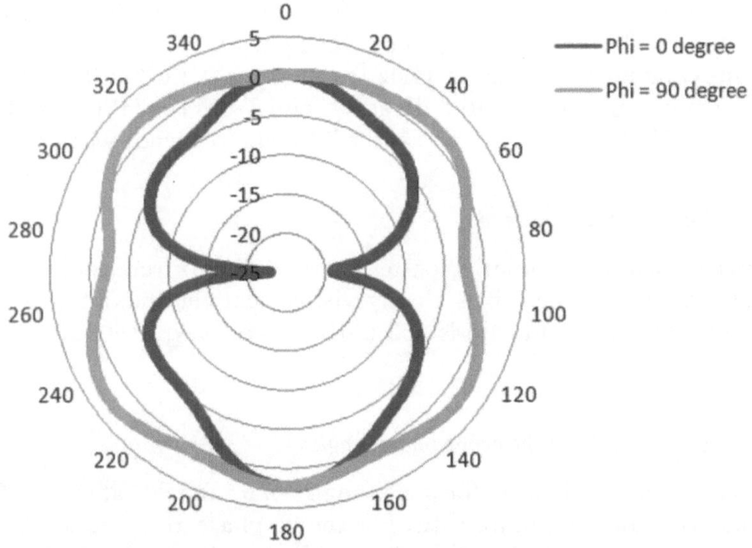

Figure 6.10 Radiation pattern at 7.1 GHz for phi = 0° and 90°.

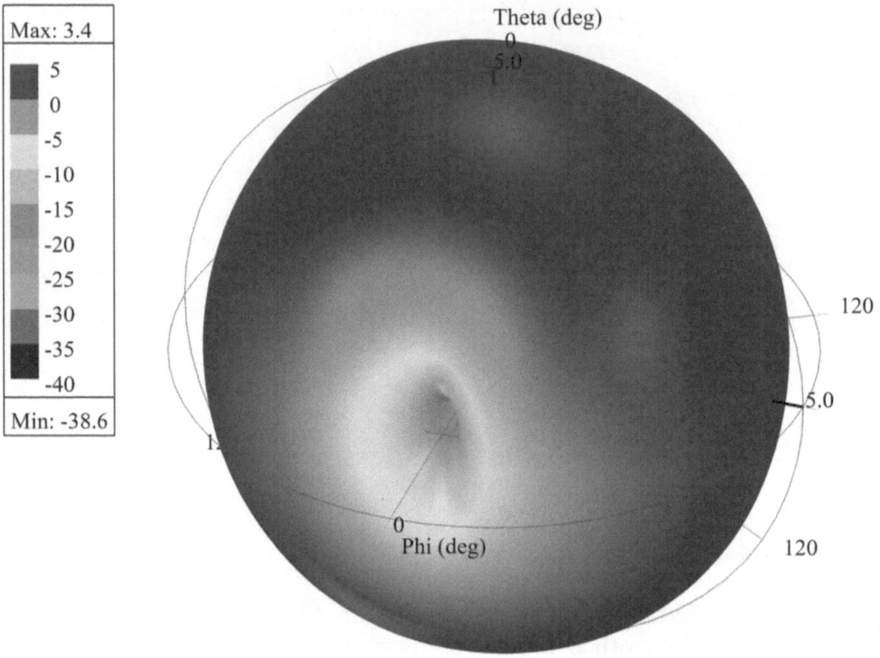

Figure 6.11 3-D radiation pattern at 4.6 GHz.

The maximum gain at the 4.6 GHz frequency is 3.4 dB at 0° and 180°, while the minimum gain is –38.6 dB at 90° and 270°. The half power beam width (HPBW) is almost 70°, and the first null beam width (FNBW) is exactly 180° at the 4.6 GHz frequency when phi = 0°.

The maximum gain at the 7.1 GHz frequency is 0.2 dB at 0° and 180°, while the minimum gain is –30.6 dB at 90° and 270°. The HPBW is almost 40°, and the FNBW is exactly 180° at the 7.1 GHz frequency when phi = 0°.

6.2.1.1.5 3-D radiation pattern

The three-dimensional radiation pattern at the 4.6 GHz frequency is shown in the Figure 6.11, where it is clearly visible that the maximum gain is 3.4 dB and the minimum gain is –38.6 dB. It is also symmetrical in every direction.

6.2.1.1.6 Electric field at different phase angles

The electric field distribution for phase angles of 0° and 90° at the 4.6 GHz frequency is shown in Figure 6.12. For the 0° phase angle, it is clear the maximum electric field is found in the feed line of the patch and the patch

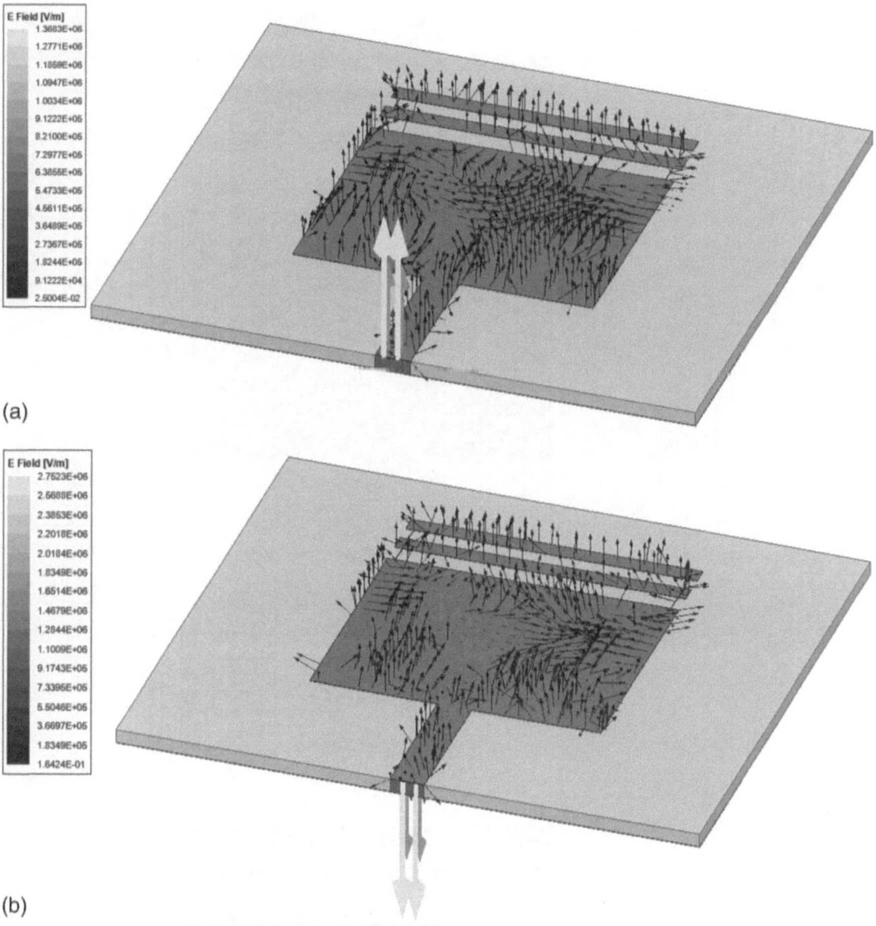

Figure 6.12 Electric field distribution at (a) a phase angle of 0° and (b) a phase angle of 90°.

has the minimum electric field distribution, but the second stub has a little bit higher electric field density compared to the patch.

For the phase angle of 90°, the result is different. In that case, the maximum electric field distribution is found at the bottom left edge after the feed line. The bottom right edge also shows good electric field distribution. Both the stubs have higher electric field distribution compared to the other edges of the patch.

6.2.1.1.7 Surface current distribution

Figure 6.13 shows the surface current distribution at the 4.6 GHz and 7.1 GHz frequencies.

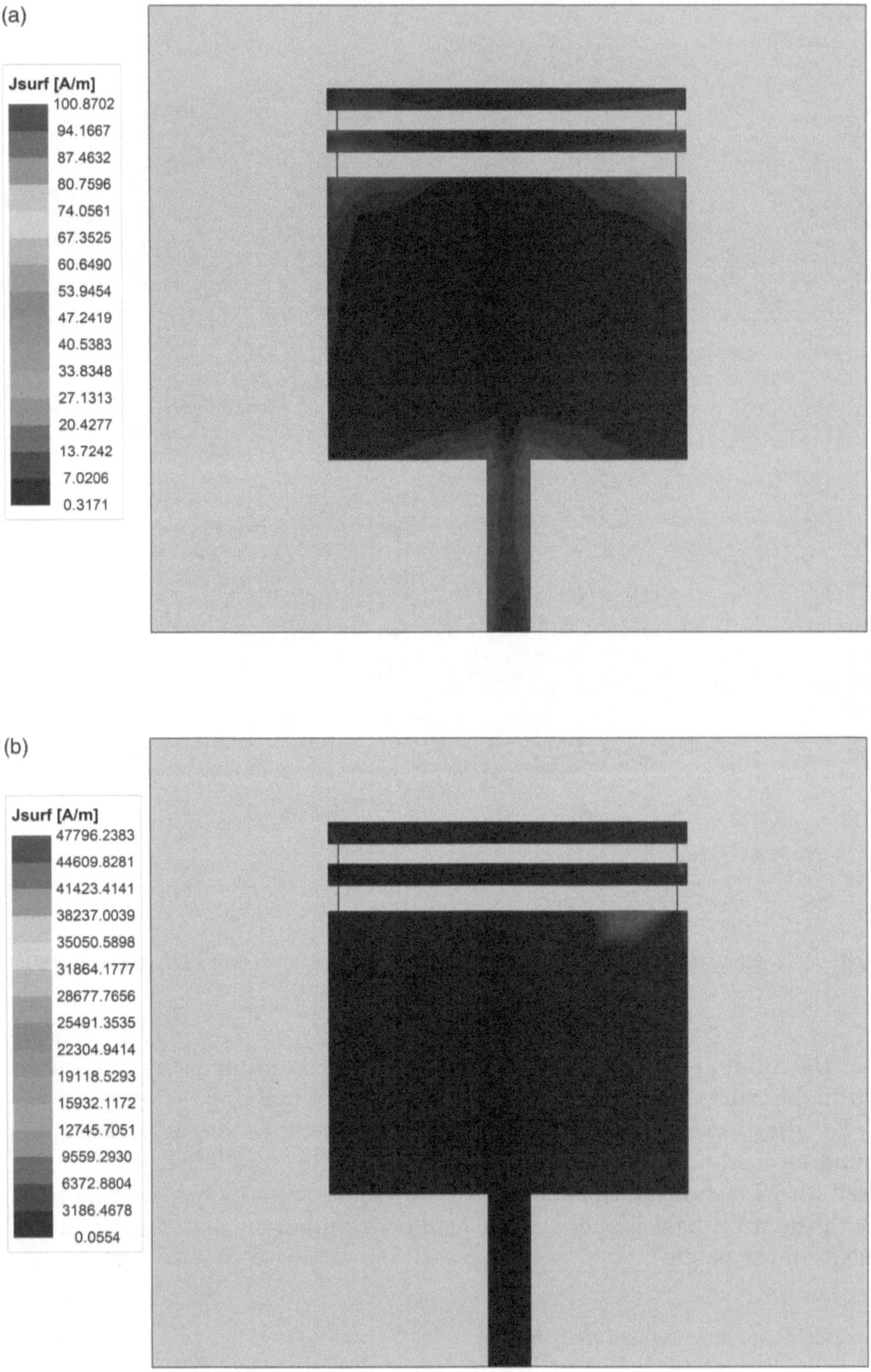

Figure 6.13 Surface current distribution at (a) 4.6 GHz and (b) 7.1 GHz.

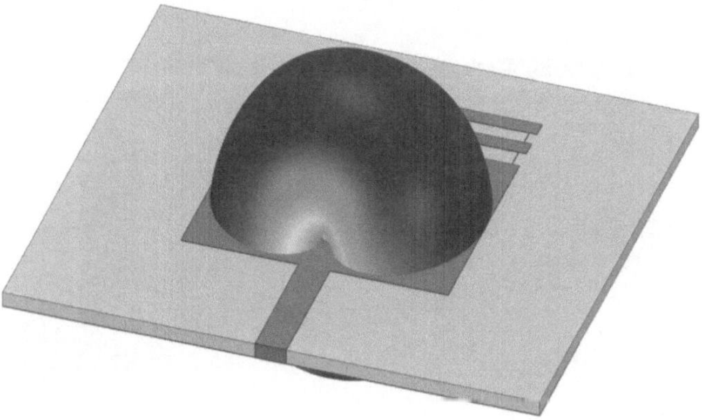

Figure 6.14 3-D radiation pattern over antenna surface at 4.6 GHz.

At the 4.6 GHz frequency, a little bit higher surface current distribution is found at the feed line of the antenna, at the top corners of the main square patch, and at the corners of the stubs compared to the entire patch. But in the case of the 7.1 GHz frequency, surface current distribution is almost the same over the entire patch except at the top right corner of the main square patch, where slightly higher surface current distribution is found.

6.2.1.1.8 3-D radiation over antenna

Figure 6.14 illustrates the 3-D radiation pattern (shown in Figure 6.11) over the antenna at the 4.6 GHz frequency.

6.2.2 Design and analysis of a UWB MIMO antenna

This section focuses on a UWB MIMO antenna [11] with a design similar to that discussed above but with two patches that are kept far apart from each other [12]. Figure 6.15 shows the top and bottom views [13] of the proposed UWB MIMO antenna, and Figure 6.16 shows a 3-D view.

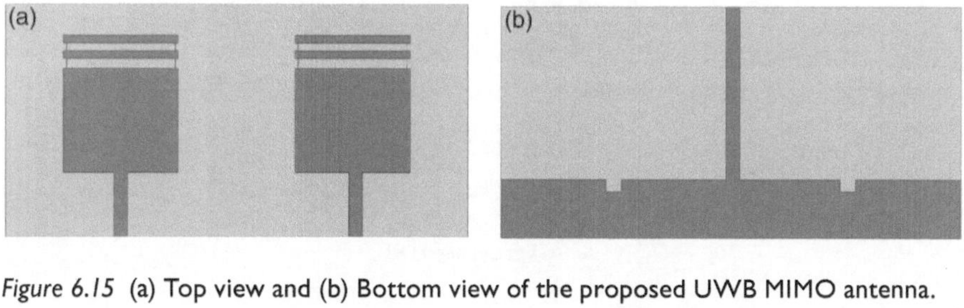

Figure 6.15 (a) Top view and (b) Bottom view of the proposed UWB MIMO antenna.

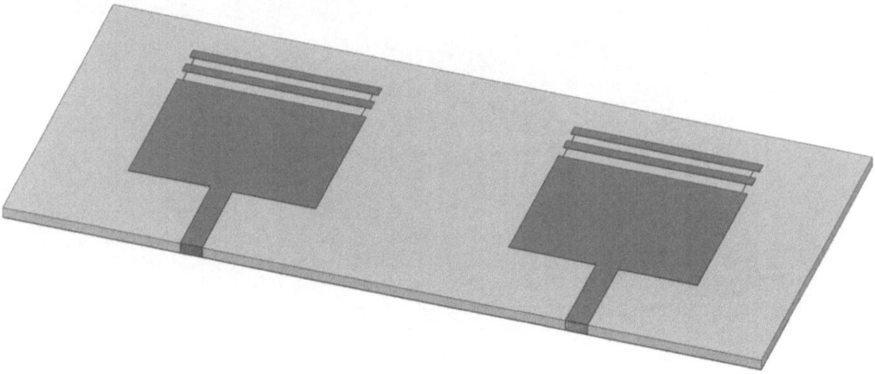

Figure 6.16 3-D view of the proposed UWB MIMO antenna.

6.2.2.1 Simulated results and analysis

The designed antenna has been simulated by ANSYS Electronics Desktop (High Frequency Structure Simulator [HFSS]) version 21.0 software. The simulated results are shown below.

6.2.2.1.1 S parameters (return loss)

Figure 6.17 shows that the two patch antennas have two resonating frequencies with band-notched characteristics [14] of 4.3–4.4 GHz and 7.6 GHz, with return losses of –15.5 dB and –10.88 dB, which are very much acceptable in the field of microwave wireless communication systems. They also provide two wide bands, one of 1.8 GHz from 3 to 4.8 GHz, and one of 0.6 GHz from 7.3 to 7.9 GHz. In these bands, the return loss is less than –10 dB

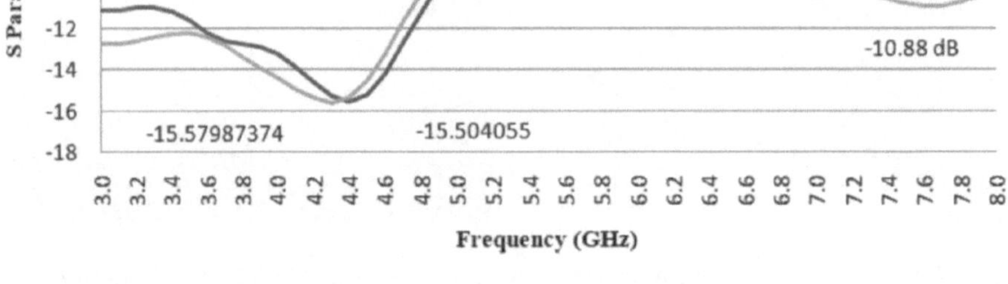

Figure 6.17 Parameters of S_{11} & S_{22}.

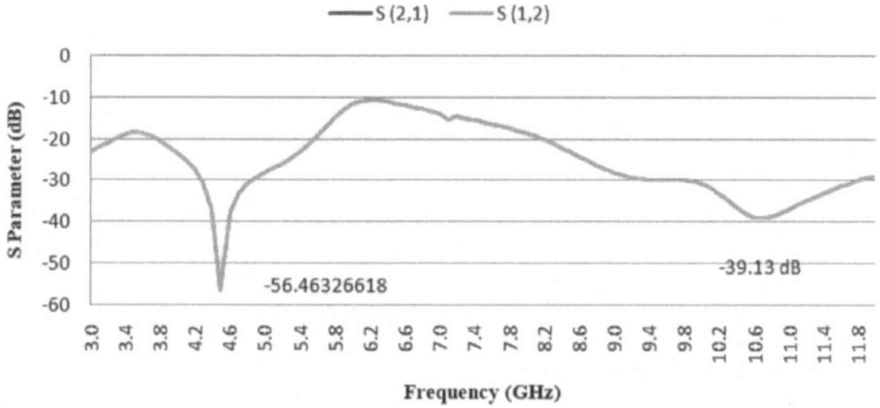

Figure 6.18 Parameters of S_{12} and S_{21}.

in the entire range. It signifies that more than 90 percent of the input power is delivered to the output power, so the power loss is very small.

Figure 6.18 shows that he mutual coupling of these antennas has two resonating frequencies of 4.5 GHz and 10.7 GHz with return losses of –56.46 dB and –39.13 dB, which are very much acceptable in the field of microwave wireless communication systems. It signifies that the mutual coupling is very small [15]. The mutual coupling [16] of the patch antennas provides a UWB of 9 GHz, ranging from 3 to 12 GHz. In this entire band, the return loss is less than –10 dB in the entire range.

6.2.2.1.2 Voltage standing wave ratio

Figure 6.19 shows that the two resonating frequencies have VSWRs of 1.4 and 1.8. The frequency bands of 1.8 GHz, ranging from 3 to 4.8 GHz, and 0.6 GHz, ranging from 7.3 to 7.9 GHz, also have VSWRs between 1 and 2. This signifies that the impedance matching is well achieved [17, 18] and is very much in the acceptable range.

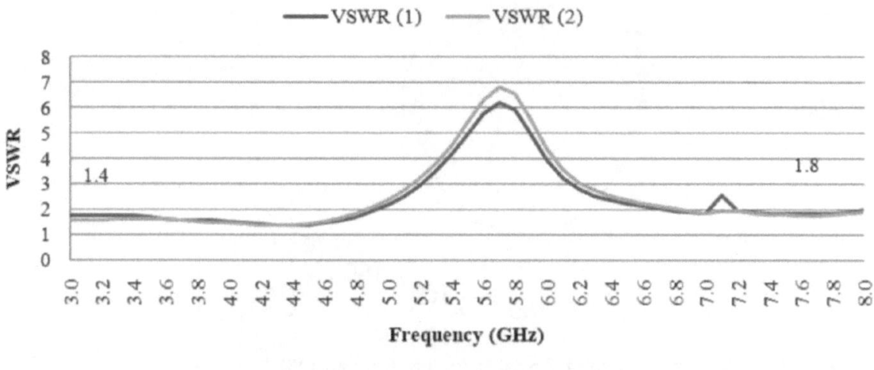

Figure 6.19 VSWR of the two patch antennas.

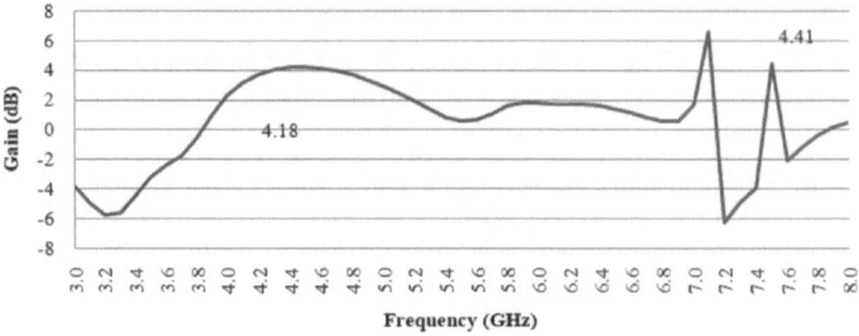

Figure 6.20 Gain vs. frequency curve.

6.2.2.1.3 Gain vs. frequency

Figure 6.20 shows that the two resonating frequencies have positive gains. The absolute value of the gain is 4.18 at 4.5 GHz and 4.41 at 7.5 GHz.

6.2.2.1.4 Radiation pattern

The radiation pattern of the proposed UWB MIMO antenna at 7.5 GHz is shown in Figure 6.21. The bidirectional characteristics can be seen when the angle phi 0°.

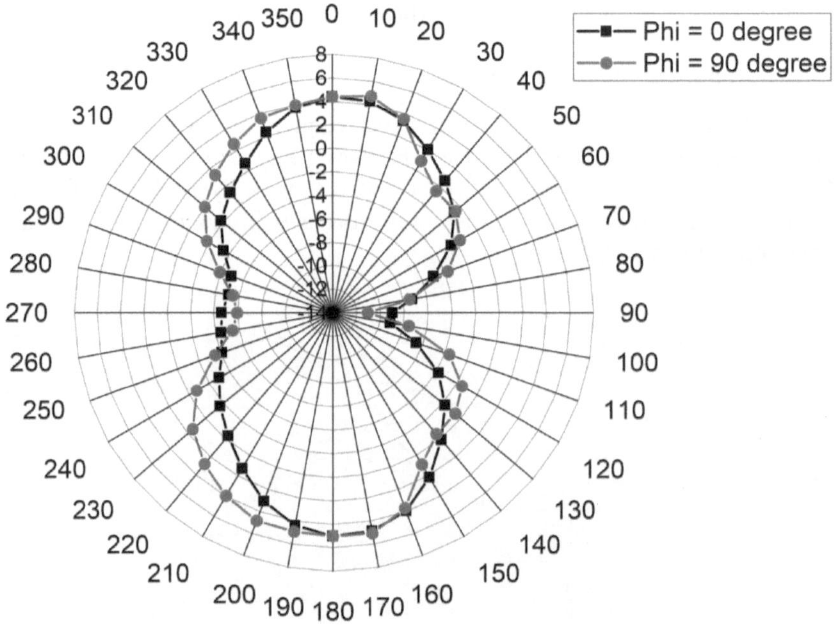

Figure 6.21 Radiation pattern at 7.5 GHz for phi = 0° and 90°.

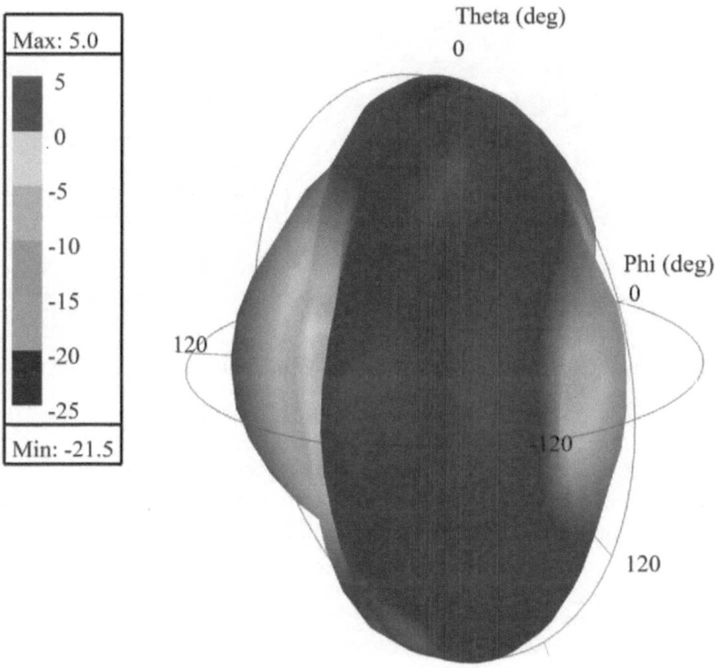

Figure 6.22 3-D radiation pattern at 7.5 GHz.

The maximum gain at the 7.5 GHz frequency is 5 dB at 0° and 180°, while the minimum gain is –21.5 dB at 90° and 270°. The HPBW is 50°, and the FNBW is exactly 180° at the 7.5 GHz frequency when phi = 0°.

6.2.2.1.5 3-D radiation pattern

The three-dimensional radiation pattern at the 7.5 GHz frequency is shown in Figure 6.22, where it is clearly visible that the maximum gain is 5 dB and the minimum gain is –21.5 dB. It is asymmetrical in nature.

6.2.2.1.6 Electric field at different phase angles

The electric field distribution for the phase angles of 0° and 90° at the 7.5 GHz frequency is shown in Figures 6.23 and 6.24, respectively. For 0° phase angle, it is clear the maximum electric field is found in the feed line of the left patch and the patch has the minimum electric field distribution, but the right stub has a little bit higher electric field density compared to the patch. The left patch does not show significant electric field distribution.

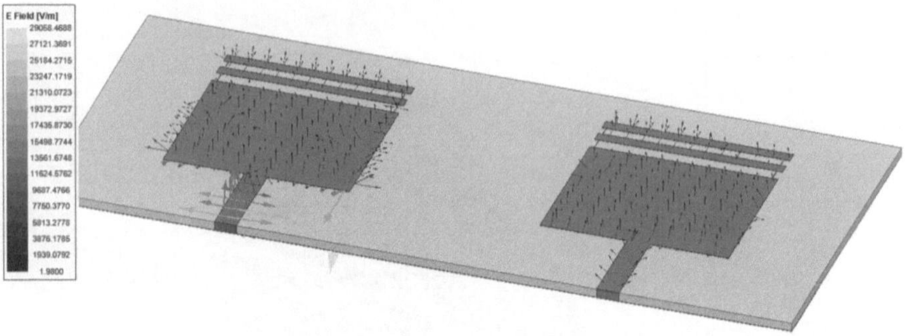

Figure 6.23 Electric field distribution at 7.5 GHz when phi = 0°.

For the phase angle of 90°, the result is different. In that case, the maximum electric field distribution is found at the same feed line, but it is much higher than the previous one. The bottom left edge also shows good electric field distribution. Both the stubs have less electric field distribution compared to the other edges of the patch.

6.2.2.1.7 Surface current distribution

Figures 6.25 and 6.26 show the surface current distribution at 7.5 GHz when phi is 0° and 90°, respectively.

At phi = 0°, a little bit higher surface current distribution is found at the feed line of the left patch antenna, at the top corners of the main square patch, and at the corners of the stubs compared to the entire patch. But in the case of phi = 90°, it is higher in the feed line compared to that of phi = 0°,

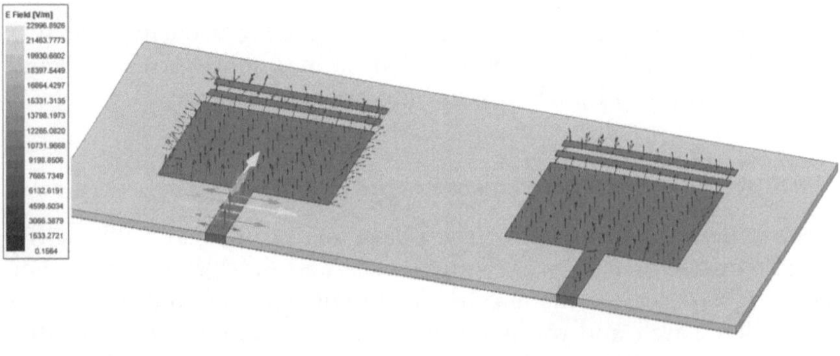

Figure 6.24 Electric field distribution at 7.5 GHz when phi = 90°.

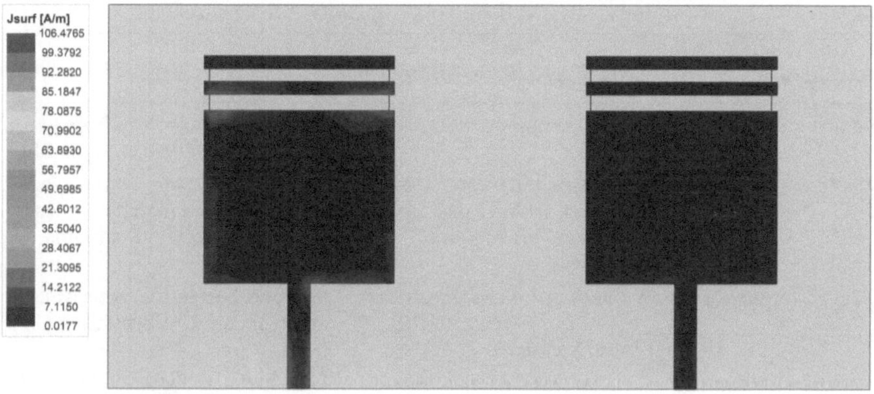

Figure 6.25 Surface current density at 7.5 GHz when phi = 0°.

while the right patch antenna shows much less surface current distribution compared to the other antenna.

6.3 COMPARISON BETWEEN UWB ANTENNA AND UWB MIMO ANTENNA

From the above characteristics of the UWB and UWB MIMO antennas, we find that larger bandwidth is produced in the UWB MIMO antenna compared to UWB antenna. All the comparisons of these two antennas are listed in Table 6.1.

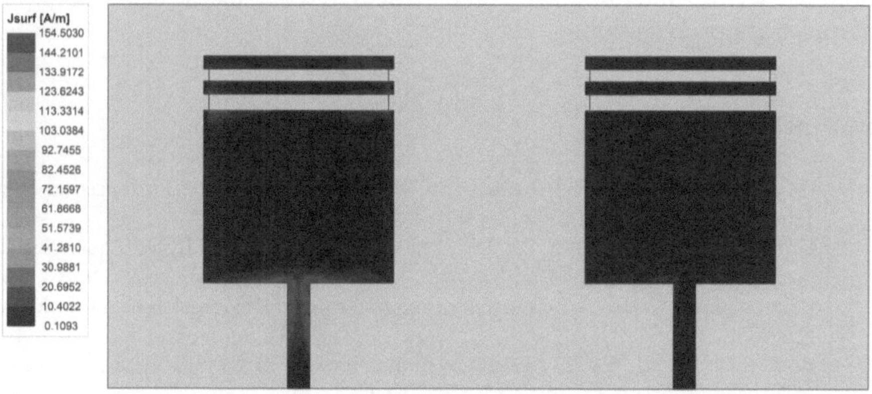

Figure 6.26 Surface current density at 7.5 GHz when phi = 90°.

Table 6.1 Comparison between UWB antenna and UWB MIMO antenna

Parameters or characteristics	*UWB antenna*	*UWB MIMO antenna*
Return loss	Lower compared to UWB MIMO.	Higher compared to UWB in S_{11} and S_{22} and lower in S_{12} and S_{21}.
VSWR	Just a little bit greater than 1 and well within range, so impedance is very well matched.	Smaller than 2 and within range, so impedance matching is less compared to UWB antenna.
Gain vs. frequency	It shows good characteristics but not as much as UWB MIMO antenna.	It shows better characteristics than simple UWB antenna [19].
Radiation pattern	Gain is lower compared to UWB MIMO antenna at resonant frequency.	Gain is much higher and better compared to UWB antenna at resonant frequency [20].
Radiation efficiency	70%–80%	> 80%

6.4 CONCLUSION

The UWB MIMO antenna has a good radiation pattern, gain with respect to the frequency, and also higher bandwidth compared to the simple UWB antenna. Based on the return loss response in S_{12} and S_{21}, it processes a huge bandwidth, whereas the UWB antenna has a smaller bandwidth compared to UWB MIMO antenna. From the research, it can be concluded that the UWB MIMO antenna provides stable gain, which is suitable for UWB wireless communication and radar communication.

Future research may focus on minimizing the area as well as optimize the surface current distribution on the patch antenna in order to avoid interference with other existing communication systems. In satellite communication, the MIMO antenna can be very efficient in reducing the disadvantages of multipath propagation.

REFERENCES

[1] Deschamps, G. A. (1953) Microstrip Microwave Antennas. Presented at the Third USAF Symposium on Antennas.

[2] Gutton, H., and Baissinot, G. (1955) "Flat Aerial for Ultra High Frequencies." French Patent No. 703 113.

[3] Balanis, C. A. (2016) Antenna Theory: Analysis and Design, 4th ed. Hoboken, NJ: Wiley.

[4] Bizaki, H. K., ed. (2011) MIMO Systems: Theory and Application. London: InTechOpen. http://dx.doi.org/10.5772/610

[5] Majeed, H., Umar, R., and Basit, A. A. (2011) Smart Antennas—MIMO, OFDM, Single Carrier FDMA for LTE. Master's thesis, Linnaeus University, Sweden.

[6] Jusoh, M., Jamlos, M. F., Kamarudin, M. R., and Malek, F. (2012) A MIMO Antenna Design Challenges for UWB Application. Progress in Electromagnetics Research B, 36, 357–371. http://dx.doi.org/10.2528/PIERB11092701

[7] Najam, A. I., Duroc, Y., and Tedjini, S. (2009) Design of MIMO Antennas for Ultra-Communications. MajecSTIC 2009, Acignon.

[8] Mietzner, J., Schober, R., Lampe, L., Gerstacker, W. H., and Hoeher, P. A. (2009) Multiple-Antenna Techniques for Wireless Communications—A Comprehensive Literature Survey. IEEE Communication Surveys and Tutorials, 11, 87–105. http://dx.doi.org/10.1109/surv.2009.090207

[9] Kaiser, T., and Zheng, F. (2010) Ultra Wide Band Systems with MIMO. Chichester: Wiley.

[10] Najam, A. I., Duroc, Y., and Tedjini, S. (2011) UWB-MIMO Antenna with Novel Stub Structure. Progress in Electromagnetics Research, 19. http://dx.doi.org/10.2528/pierc10121101

[11] Sibille, A., Oestges, C., and Zanella, A. (2011) MIMO: From Theory to Implementation. Burlington, MA: Academic Press.

[12] Fang, D. G. (2010) Antenna Theory and Microstrip Antennas. New York: CRC Press.

[13] Ibrahim, A. A., Abdalla, M. A., Abdel-Rahman, A. B., and Hamed, H. F. (2014) Compact MIMO Antenna with Optimized Mutual Coupling Reduction Using DGS. International Journal of Microwave and Wireless Technologies, 6, 173–180.

[14] Zhao, Y., Zhang, F. S., Cao, L. X., and Li, D. H. (2019) A Compact Dual Band-Notched MIMO Diversity Antenna for UWB Wireless Applications. Progress in Electromagnetics Research, 89, 161–169.

[15] Valderas, D., Crespo, P., and Ling, C. (2010) UWB Portable Printed Monopole Array Design for MIMO Communications. Microwave and Optical Technology Letters, 52.

[16] Toktas, A., and Akdagli, A. (2015) Compact Multiple-Input Multiple-Output Antenna with Low Correlation for Ultra-Wide-Band Applications. IET Microwaves, Antennas & Propagation, 9, 822–829.

[17] Azarm, B., Nourinia, J., Ghobadi, C., Majidzadeh, M., and Hatami, N. (2018) A Compact WiMAX Band-Notched UWB MIMO Antenna with High Isolation. Radioengineering, 27, 983–989.

[18] Li, W. T., Hei, Y. Q., Subbaraman, H., Shi, X. W., and Chen, R. T. (2016) Novel Printed Filtenna with Dual Notches and Good Out-of-Band Characteristics for UWB-MIMO Applications. IEEE Microwave and Wireless Components Letters, 26, 765–767.

[19] Ngoc Lan Nguyen, Van Yem Vu (2019) Gain enhancement for MIMO antenna using metamaterial structure, International Journal of Microwave and Wireless Technologies, Volume 11, Issue 8, October 2019, pp. 851–862. DOI: https://doi.org/10.1017/S175907871900059X

[20] Singh, R., & Karia, D. (2017). High gain ultra wide band MIMO antenna. 2017 2nd International Conference for Convergence in Technology (I2CT). DOI: https://doi.org/10.1109/i2ct.2017.8226224

Chapter 7

Four-element wave patch multiband MIMO antenna for 5G application

Richa Kumari, Yadwinder Kumar, Ajay Mudgil, and Balwinder Singh

CONTENTS

7.1 INTRODUCTION

Wireless communication relies on the efficient working of antennas. For today's wireless devices, miniaturization of antenna sis emphasized. Patch antennas are found to have a huge role in this evolving era of technology. They are manufactured by the process of photolithography; that is, they are etched on a printed circuit board (PCB). Patch antennas are known as the most lightweight, low-cost, easy-to-fabricate, and low-profile antennas. However, they have some drawbacks like narrow bandwidth, efficiency lag, and lesser directivity as opposed to many other antenna types. Hence, they are used in an array to overcome such disadvantages. Multiple-input

DOI: 10.1201/9781003290230-7

multiple-output (MIMO) antenna technology has paved the way for high data rate and least fading requirements. MIMO allows several antenna elements to operate together simultaneously establishing multiple parallel channels on the same frequency band and at the same total radiated power to exploit multipath propagation.

The two methods employed for MIMO wireless communication are space-time block coding and spatial multiplexing. In space-time block coding, a copy of the same data that has been transmitted at time t_1 is transmitted again at time t_2. However, this is not an exact copy but a manipulated one, so that the exact information can be extracted out of the multipath propagation. It does not enhance the bit rate but is useful in reducing the bit error rate. In spatial multiplexing, data is split and transmitted via different antenna elements to the receiver, where, after signal processing, the exact data is retrieved. Its disadvantages include the complex signal processing required after signal reception. Both the techniques are used together in wireless systems to gain benefits in terms of both coverage and data rate, according to modified Shannon equation, shown in Equation 7.1 [1]. This technique has been widely adopted in the latest wireless standards such as wireless local area network (WLAN), Worldwide Interoperability for Microwave Access (WiMAX), and Long-Term Evolution (LTE).

$$C = MB\log\left(1+\frac{N}{M}\times SNR\right) \tag{7.1}$$

where M is the number of antennas on the transmitter side, B is the bandwidth in Hertz, C is the channel capacity in bits per second, N is the number of antennas on the receiver side, and SNR is the signal-to-noise ratio.

The channel capacity increases with the increase in antenna elements at the transmitting and receiving ends. The use of multiple antenna elements at the transmitter and receiver has increased since its introduction by Arogya Swamy Paul Raj and Thomas Kailath in 1993. They proposed the method of spatial multiplexing by using MIMO technology. MIMO relies on the use of multiple antenna elements at the base transceiver station and the user end. The base transceiver station has enough space to accommodate several antenna elements at the requisite distance from each other. However, user-end terminals such as mobile phones and routers have less space for accommodating various antenna elements. When these elements are too close in proximity, it leads to high mutual coupling and channel correlation, which further tend to degrade antenna performance. This coupling between antenna elements should be addressed to enhance channel capacity. Further, channel correlation and mutual coupling can be reduced by the distinct radiation patterns of antenna elements placed close to each other.

The user end makes use of the patch antenna. A patch antenna is low profile, lightweight, low cost, and easy to manufacture and is configured with enough technology to work with recent wireless communication systems. Often called

PCB antennas, they are etched on a PCB using the process of photolithography. A patch antenna has a substrate, a patch, a feed line, a feed, and a ground, as shown in Figure 7.1. On providing the feed, the region above the ground and the region below patch are ionized to opposite kinds of charges, creating attractive and repulsive forces. The attractive force exists between opposite charges above the ground plane and below the patch, whereas the repulsive force exists among the like charges below the patch. As a result of this, some charges move up the patch. For $h/W \ll 1$ (h is the height of the substrate and W is the width of the patch), the attractive force is dominant over the repulsive one. But since there is some current at the edges of the patch because of the transfer of some charges though the dielectric and ground plane which is not truncated to the length and breadth of the patch but extended beyond it. There is fringing shown by the front and end terminals of the patch to the dielectric. The rest of the terminals have their fields canceled out.

Fringing is often responsible for the frequency of the radiated waves generated and many other antenna parameters [2]. The fringing field of the patch antenna has a devastating effect on the mutual coupling between closely spaced antenna elements [3]. Also, the dielectric acts in different modes in an operation. These modes signify the field distribution in the dielectric material. The deciding factors of the mode are the length and width of the patch.

Several patch elements in an antenna are used together in the same frequency band and at the same total radiated power but with different port terminals to form a MIMO. They are designed with enough technology to support recent wireless communication. MIMO antenna parameters include the total active reflection coefficient, isolation, correlation coefficient, mean

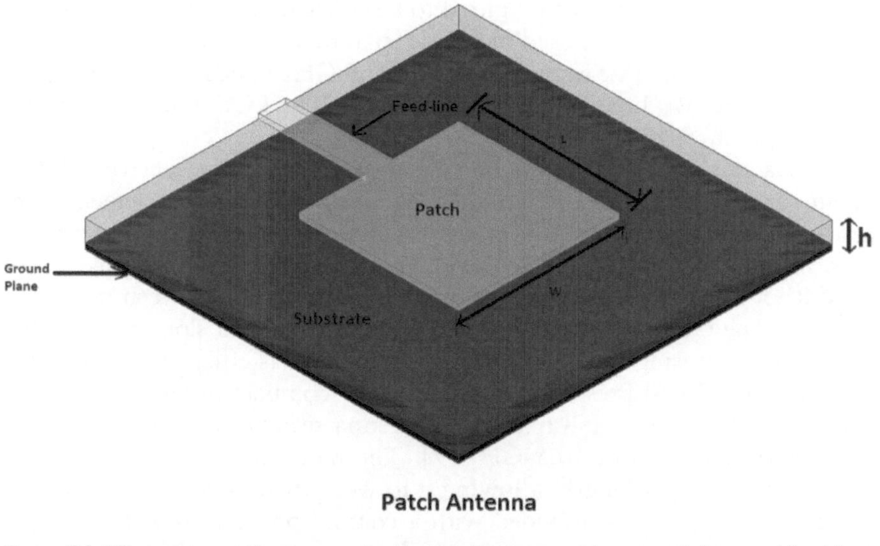

Figure 7.1 Microstrip patch antenna showing a substrate, patch, ground plane and feed line.

effective gain, efficiency, and diversity gain. When designing a MIMO antenna, these parameters have to be taken into consideration. Before proceeding to fabricate an antenna, these parameters are brought into the picture via the use of electromagnetic simulation software. Various antennas and their performance matrices are discussed below.

Han and Choi (2011) presented a MIMO antenna with two dipole antenna elements placed on two opposite sides of the FR4 substrate [4]. The antenna is made to resonate in WLAN (698–812 MHz) and mobile WiMAX (M-WiMAX) (2.12–2.84 GHz) bands. The two antenna elements have been arranged orthogonally with respect to each other with a coaxial feed at the center. The two elements have been separated by a distance of 0.8 mm and have good isolation of more than 20 dB. Good isolation on account of polarization diversity is achieved for the multiband MIMO antenna. An ECC value of less than 0.3 is achieved in bands of interest.

Han and Choi (2010) presented a MIMO antenna with two plantar inverted-F antenna (PIFA) elements [5]. The PIFAs are accompanied by slits as feeding and shorting lines. The MIMO antenna has a band matching circuit at one end to make the antenna work at LTE bands. Further by changing the inductance value or the width of the slits, the resonant value of the antenna can be tuned independently. The two PIFAs are placed at two ends to resonate at three bands—LTE, Wideband Code Division Multiple Access (W-CDMA), and M-WiMAX—so the MIMO antenna is well suited for the fourth generation (4G) mobile system.

Su (2010) presented a MIMO antenna design in which three antenna elements are placed on a foam-based ground plane (a copper-nickel-zinc alloy was used here for the prototype) in a rotating arrangement with equal angles of inclination to each other [3]. The proposed antenna, compiled for an internal access point, is configured to be concealed in a casing of height 10 mm. The design, composed of two loop antennas with a common feed point, is made to operate in the 2.5 and 5.2/5.8 GHz bands with a high gain of 7 dBi in the two bands with good isolation. The ECC values are below 0.007 in the three bands.

Malviya et al. (2018) proposed a design of a 2 × 2 trans-receive MIMO antenna to operate in five LTE bands [6]. Two two-antenna elements are arranged on two smaller lengths of an FR4 dielectric substrate of dimensions 68 × 98 mm^2. The MIMO antenna exhibits a voltage standing wave ratio (VSWR) of less than 2 and more than 10 dB of isolation between radiating elements. The C-shaped radiating element with C-shaped slot proved to be effective in increasing the current path. The design has efficient ECC values.

Kaur et al. (2016) presented a miniaturized compact antenna design of dimensions 25 × 25 × 1.58 mm^3. The antenna resonates at five frequency bands between 6.6 and 10.5 GHz [7]. The patch has three rectangular slots along its longer length, allowing it to work differently than a normal rectangular shape. It is provided with a coaxial probe feed at the end of the patch at a position providing good adaptation to resonant frequencies.

Kaur et al. (2017) presented a compact miniaturized antenna containing multiple C-shaped slots of varying sizes in the patch. [8]. It has a partial ground plane and microstrip line feed. The antenna resonates at three frequency bands. The radiation pattern obtained is omnidirectional, exhibiting valuable gain over the required frequency bands.

Sharawi et al. (2013) presented an isolation enhancement method by introducing capacitively loaded loops (CLLs) in the antenna [9]. The CLL and complementary CLL elements have been employed on the top and bottom sides of the substrate. The CLL on the top side is placed on the patch between two 4-shaped patch elements, while on the other side the complementary CLL is etched between the two regions opposite the patches. The resonance obtained out of the CLL—i.e., the metamaterial unit element (UE)—depends on the capacitance and inductance associated with the spiral structure. Reducing the length of the spiral structure results in more gap between its edges and hence a reduction in the shape of the UE, which also tends to resonate at a higher frequency band, and vice versa. The shorter CLL is employed on the top side of the substrate, while the longer one is employed on the bottom side. The bands of resonance were found to be 827 to 853 MHz and 2.3 to 2.98 GHz with isolation of more than 10 dB and 3 dB, respectively.

Zhang et al. (2012) proposed a two-element ultra-wideband (UWB) MIMO antenna [10]. The antenna comprises monopole and half-slot antennas. The slot is responsible for isolation in the lower frequency band and provides additional bandwidth to one antenna resonance region. The mutual coupling is reduced because of the distinct radiation pattern and polarization of the two antenna elements. The antenna has a dimension of 25×40 mm for dongle applications. The antenna can cover the lower UWB band from 3.1 to 5.12 GHz with isolation higher than 26 dB.

Zhou et al. (2012) discussed a MIMO antenna of two loop elements arranged one after the other [11]. The surface current distributions get concentrated in the outer loop, while resonating in the lower band and get concentrated in the inner loop while resonating for the higher band. The loops are open at a point, providing two current paths in the same loop. The longer monopole is expected to create resonance at the lower frequency of the upper band, and the shorter monopole is expected to create resonance at the higher frequency of the lower band. A U-shaped etching in the ground plane has been provided to improve isolation between the radiating elements. The antenna occupies a space of $50 \times 17 \times 0.8$ mm^3 and has an ECC value of less than 0.01 in the desired bands.

Kulkarni and Sharma (2013) presented a MIMO slot antenna and a loop antenna for use in digital TV applications as well as LTE operations [12]. The arrangement is made on an FR4 substrate. It has ECC values of less than 0.22 for both bands. Yang et al. (2010) presented a MIMO three-port antenna [13]. The antenna is composed of a loop antenna and PIFAs with self-complementary structures that resonate at the 5 band and 2 GHz bands. The antenna also resonates at 370–870 MHz (ultrahigh frequency or

UHF band) and 1300–1520 MHz (L band). Fernandez and Sharma (2013) presented a meandered loop MIMO antenna [14]. It has a 1 × 2 array configuration. Its coverage bands lie in the region of 4G LTE, Universal Mobile Telecommunications Service (UMTS), WiMAX, and WLAN. Results show an omnidirectional pattern of radiation for the antenna.

Foudazi et al. (2011) presented a dual-band wide-slot antenna structure [15]. The antenna initially is capable of resonating at 3.1–10.6 GHz (UWB band) with a U-shaped patch structure, but with the introduction of a T-shaped monopole path, it resonates in one more band corresponding to 2.4 GHz (WLAN band). The design is printed on an FR4 substrate with a conducting ground plane that has a pentagonal section removed from it. This reduction supported an omnidirectional radiation pattern.

Han and Choi (2011) presented a MIMO antenna with two PIFA elements that have an asymmetric slotted strip between them [16]. The slotted structure helps improve the isolation characteristics between the LTE and Wireless Fidelity (Wi-Fi) bands. A jointed shorting line is provided to reduce interaction between the two PIFA elements. Cheon et al. (2012) compared the results when ferrite material was used as an antenna substrate in place of FR4 material [17]. Both materials used are of the same permittivity. The magneto-dielectric material allows more miniaturization and impedance bandwidth of the antenna as compared to the FR4 substrate. However, the magneto-dielectric material is less efficient because of magnetic loss.

Hong et al. (2008) presented a MIMO antenna operating in the UWB range (2.27 to 10.2 GHz) [18]. The antenna is composed of two Y-shaped radiating patches on a thin dielectric substrate of height 0.8 mm. Three stubs are inserted in the ground plane to reduce coupling between the two radiating elements. The presence of the stubs makes a remarkable difference in the surface current distribution in the radiators and on the ground plane.

Kumar and Singh (2017) proposed a fractal antenna of heptaband resonant behavior [19]. The ground plane has an L-shaped structure, which helps to provide a higher gain value. Three iterations of the antenna have been developed: the first iteration supports three resonant bands, the second supports five resonant bands, and the third supports seven resonant bands. The bands of resonance are at 2.4, 4.437, 5.38, 7.01, 7.6, 8.41, and 9.09 GHz. Also the VSWR values lie in the valuable range for the bands of interest.

7.2 ANTENNA CONFIGURATION

The design proposed in this chapter is shown in Figure 7.2. It has four identical antenna elements that have been placed orthogonally with respect to each other. Each element has a length of 20 mm and a width of 36.45 mm (greater than half the length of the substrate). The four elements have been placed on a square FR4 dielectric substrate whose permittivity (ε_r) is 4.4 and loss tangent (δ) is 0.02. The dimensions chosen for the substrate are $70 \times 70 \times 1.58$ mm^3. The elements have been provided with a spatial arrangement on the square surface. The four radiators have been placed orthogonally with respect to each other but have not been aligned parallel with the edge of the substrate, as shown in Figure 7.2. The length of the patch has been tilt of 5.67° with respect to the horizontal. The ground of each element is placed in exactly the same position and alignment as the corresponding patch element. The design is based on the radiation mechanism of electromagnetic waves through an antenna. A circular slot of radius 5 mm is cut at the center of the substrate, as shown in Figure 7.2. The figure shows the top, bottom, and isometric views of the antenna.

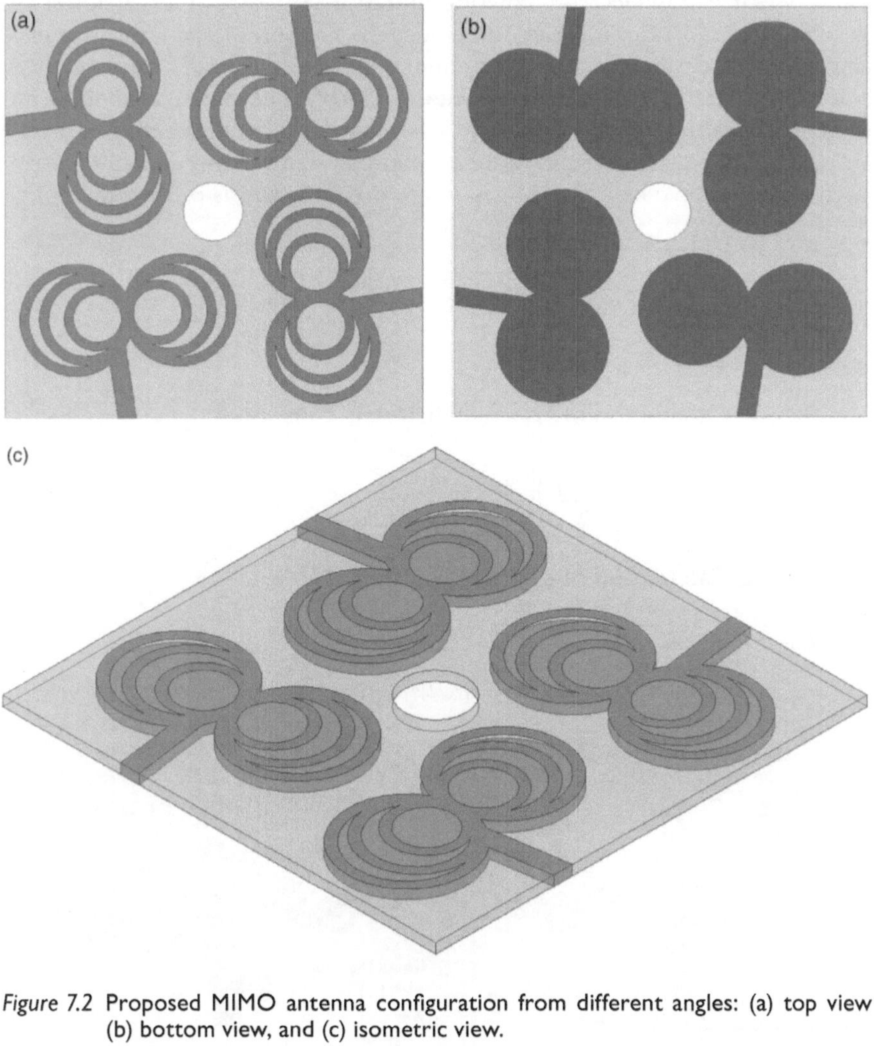

Figure 7.2 Proposed MIMO antenna configuration from different angles: (a) top view (b) bottom view, and (c) isometric view.

The geometry of the patch of the proposed MIMO antenna element is shown in Figure 7.3(a). Circular strips with radii of 5.5, 7.5, and 9.5 mm are used. Each strip has inner and outer circles of the same or different dimension. The innermost strip has an inner circle with a shift in the center of 0.5 mm toward the outer strip from the original position, and the outer circle has a shift in the center of 0.375 mm and a shift in the radius of 0.125 mm toward the outer strip from the original position. The second strip has center and radius shifts of 0.25 mm and 0.25 mm, respectively, in both the circles toward the outer strip from its original position. The last strip, of radius 9.5 mm, has no shifts at all. The *original position* term is used to signify the distance from the center of longer side of the patch. Also, the three strips are joined to form a structure. The entire work of simulation is performed in electromagnetic simulation software. The thickness of each strip is 1.55 mm. On the other side, the ground of each antenna element has been designed with two circular shapes of radius 9.5 mm attached to form a structure, as shown in Figure 7.3(b). The patch resembles an off-centric multiple rings structure, whereas the ground resembles a figure eight–shaped structure. Deduced equations of a rectangular microstrip patch antenna have been used to estimate the dimensions of the patch [2]:

- Step 1: Calculation of width

$$w = \frac{c}{2f_0\sqrt{\frac{\varepsilon_r + 1}{2}}} \tag{7.2}$$

- Step 2: Calculation of effective dielectric constant (ε_{reff})

$$\varepsilon_{reff} = \left(\frac{\varepsilon_r + 1}{2}\right) + \left(\frac{\varepsilon_r - 1}{2}\right)\left[1 + 12\left(\frac{h}{w}\right)\right]^{\frac{-1}{2}} \tag{7.3}$$

- Step 3: Calculation of effective length (L_{eff})

$$L_{eff} = \frac{c}{2f_0(\varepsilon reff)^{1/2}} \tag{7.4}$$

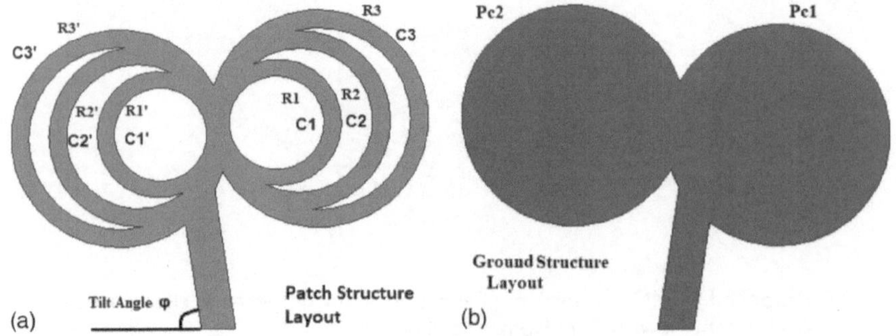

Figure 7.3 Layout of proposed geometry: (a) patch structure and (b) ground structure.

- Step 4: Calculation of length extension (ΔL)

$$\frac{\Delta L}{h} = 0.412 \frac{(\varepsilon_r + 0.3)\left(\frac{w}{h} + 0.264\right)}{(\varepsilon_r - 0.258)\left(\frac{w}{h} + 0.8\right)} \tag{7.5}$$

- Step 5: Calculation of actual patch length (L)

$$L = L_{\text{eff}} - 2\Delta L \tag{7.6}$$

The width and length of the antenna obtained after calculation at $f = 3$ GHz are 30.42 and 23.44 mm. The patch width and length are 36.45 and 20 mm, which are very much in agreement with the calculated values, though it is a fractal and circular strips structure. The feed provided to the radiators is a microstrip line feed. Its thickness is 3 mm, which provided a good match for the antenna element corresponding to a 50 Ω coaxial SMA connector. The feed line is tilted with respect to the vertical by 8.5°. This tilt causes the feed line to meet the antenna on one side of the patch rather than at the center. It is because of the tilt that it touches the antenna element at some other point other than the center of the antenna.

Table 7.1 lists the parametric details of the patch and ground of the proposed MIMO antenna.

The slot in the center of the substrate reduces the surface current excitation produced in the antenna. Hence, distortion gets reduced. The tilt forces the antenna to align to the center of another patch element. This tilt has a major impact on the fringing fields of the antenna. Also, it helps in getting an efficient radiation pattern at each resonant band. An increment in the width of the feed line can provide better results at lower frequency bands but can

Table 7.1 Details of all geometrical parameters of the proposed antenna structure

S. no.	*Parameter*	*Value*
1	R1 = R1′ – radius of inner strip on either side	4.075 mm
2	R2 = R2′ – radius of middle strip on either side	6.3 mm
3	R3 = R3′ – radius of outer strip on either side	7.95 mm
4	Thickness of strips C1, C1′, C2, C2′, C3, and C3′	1.55 mm
5	Tilt angle of patch and ground	5.67°
6	Thickness of feed line	3 mm
7	Height of feed line	11-12 mm
8	Tilt angle Φ	8.5°
9	Substrate thickness	1.58 mm
17	Substrate length	70 mm
18	Substrate width	70 mm
19	Radius of ground—circle Pc1	9.5 mm
20	Radius of ground—circle Pc2	9.5 mm

degrade results for high bands of resonance. The width of the strip in the patch is responsible for the bandwidth of resonance. Strips, when brought closer, give resonance in much closer bands. The bandgap in this case lowers to such a value that the entire band of resonance counts for one wider band. A haphazard value of return loss is obtained in that case. The outer diameter of the patch is the same as that of the ground, and the alignments are also same, leading to less fringing. Providing a small circular slot in the ground of the antenna element near the center can provide better return loss for the lower band but also can diminish the result for the higher band. The outer strip is responsible for resonance at the lower bands, and the inner strips are responsible for resonance at the higher bands. The feed line has a thickness of 3 mm, which provides good matching for the antenna near the 5 GHz band.

7.3 RESULTS AND DISCUSSION

7.3.1 Radiation regions

The radiation zone is the zone of formation of the radiation lobe of the antenna. The antenna is found to radiate well in the far-field region. Before the far-field region, the radiation lobes are found to couple with various metallic, nonmetallic, and metalloid substances and get disrupted. For the proposed antenna, the reactive near-field region is found to be less than 0.062 mm; beyond that is the radiative near-field region, which is found to be less than 0.2 mm; beyond that is the far-field region. The far-field region has the major radiation obtained by the antenna.

7.3.2 Return loss (S_{11})

The S_{11} parameter of an antenna is an important aspect that has to be addressed while designing the antenna. S_{11}, or the reflection coefficient (Γ), signifies the amount of signal that is getting reflected from the connector or terminal of the antenna. The antenna needs to be properly matched before its operation. The antenna impedance should be about 50 Ω, which has been chosen as a standard value of impedance. The return loss of the antenna depends entirely on the matching of the antenna. The reflection coefficient describes the number of electromagnetic waves reflected by an impedance discontinuity in the transmission line. Equations 7.7 and 7.8 show the relation between the reflection coefficient and the forward and reflected power, as well as voltage of the antenna [2].

$$\Gamma = \sqrt{\frac{P_{\text{ref}}}{P_{\text{fwd}}}} \tag{7.7}$$

$$\Gamma = \frac{V_{\text{ref}}}{V_{\text{fwd}}} \tag{7.8}$$

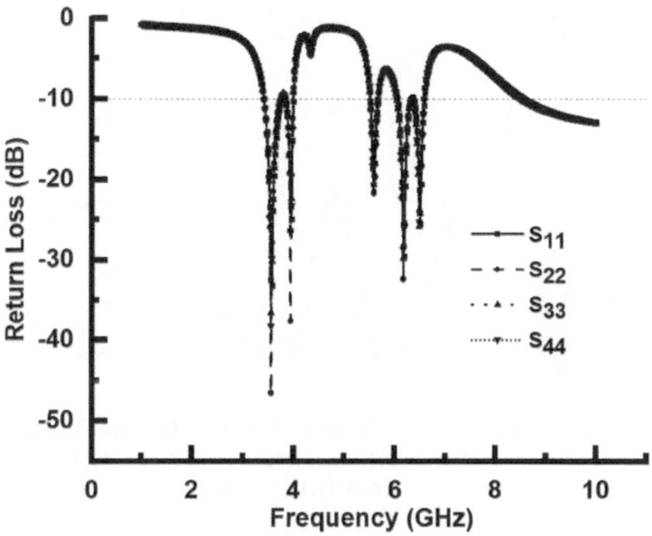

Figure 7.4 Return loss of proposed MIMO antenna.

where P_{ref} is reflected power, V_{ref} is reflected voltage, P_{fwd} is forward power, and V_{fwd} is forward voltage. Return loss is the power term of the reflection coefficient—i.e., –20 log (Reflection coefficient). The value of S_{11} is negative to assure the positive value of the reflection coefficient. Figure 7.4 shows the S_{11} parameter plot of the four antenna elements of the proposed MIMO antenna.

The resonance frequency is the point where the best matching or resonance of the RLC component of the antenna circuit is obtained. Since this is a multiband MIMO antenna, three bands of resonance have been obtained between 1 to 7 GHz. The antenna is even found to resonate in wideband near 10 GHz. The four plots can be seen to overlap. The antenna resonates in three frequency bands corresponding to 3.42 to 3.98 GHz, 5.52 to 5.66 GHz, and 6.04 to 6.58 GHz. The peak resonance has been obtained at 3.56 GHz, 3.94 GHz, 5.58 GHz, 6.18 GHz, and 6.5 GHz. The bandwidths obtained are 580, 140, and 540 MHz for the three bands, respectively.

7.3.3 Voltage standing wave ratio

The VSWR describes the relationship of maximum to minimum voltage or current waves in the transmission line of an antenna. A proper match between input impedance and antenna impedance is also an important aspect. The desired value of S_{11} as high a negative value as possible corresponding to the lower value of the reflection coefficient (Γ). The S_{11} value of –10 dB corresponds to a VSWR of 2:1, which was chosen as a standard for antenna operation. Values below –10 dB are all acceptable. A VSWR of 0:0 is the ideal case of matching in an antenna, but it is not achievable in

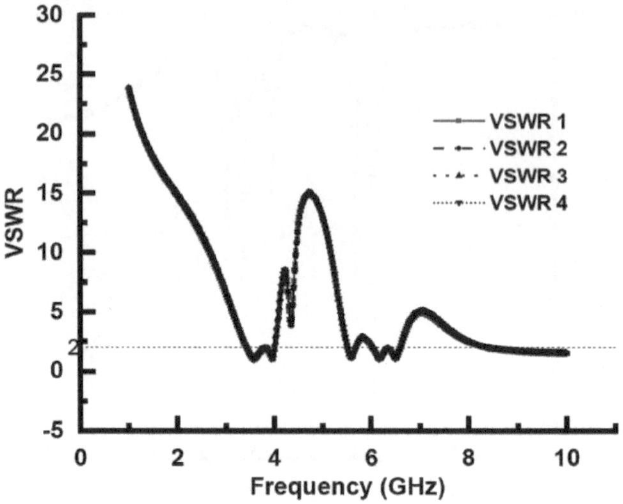

Figure 7.5 VSWR of proposed MIMO antenna.

practice. The VSWR plot obtained for the proposed antenna is shown in Figure 7.5. Equation 7.9 shows the relation between VSWR and Γ.

$$\text{VSWR} = \frac{1+\Gamma}{1-\Gamma} \tag{7.9}$$

The antenna resonating bands lie below VSWR = 2, as shown in Figure 7.5. Table 7.2 lists the resonant value ranges, peak values of resonance, and corresponding VSWR values of the proposed MIMO antenna.

7.3.4 Two-dimensional radiation pattern

7.3.4.1 E-field radiation pattern

Figure 7.6 shows the two-dimensional E-field radiation plots of the proposed antenna in the resonant bands. The E-field plots signify the E-field pattern in the elevation plane (that is, the plane perpendicular to the plane of the antenna) at θ = 0° and 90°. Both plots for each band come out almost the same. Moreover, the plots are more or less figure eight-shaped plots of the E-field radiation patterns.

Table 7.2 Return loss, bandwidth, and VSWR values of proposed MIMO antenna

Resonance frequency range (GHz)	*Peak resonance frequency (GHz)*	*Bandwidth (MHz)*	*Return loss (GHz)*	*VSWR*
3.42—3.98	3.56, 3.94	560	−38.445, −28.38	1, 1.02
5.52—5.66	5.58	140	−21.11	1.17
6.04—6.58	6.18, 6.5	540	−26.51, −24.835	1.06, 1.11

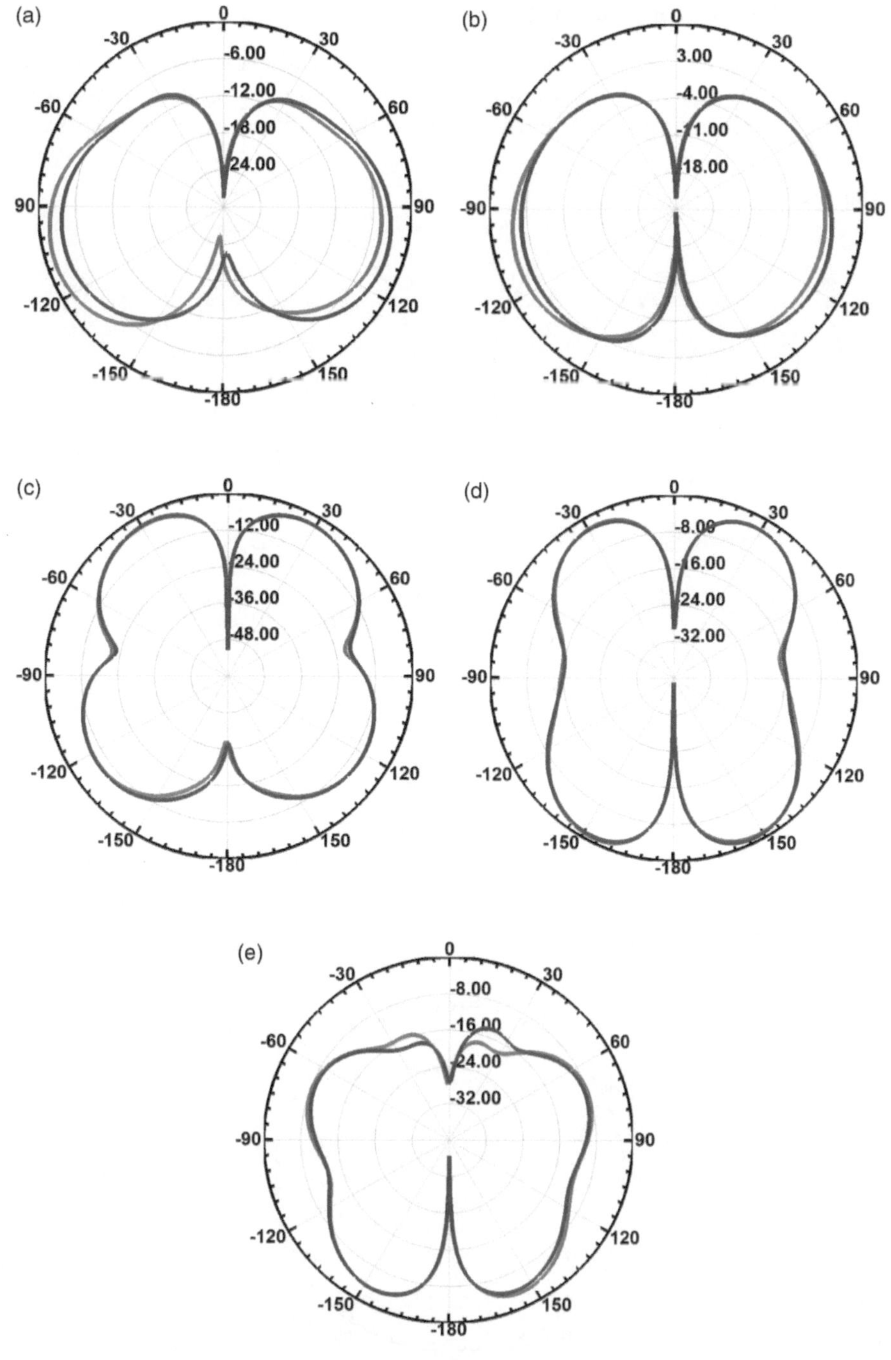

Figure 7.6 E-field radiation patterns at (a) $f = 3.56$ GHz; (b) $f = 3.94$ GHz; (c) $f = 5.58$ GHz; (d) $f = 6.18$ GHz; and (e) $f = 6.5$ GHz.

7.3.4.2 H-field radiation pattern

Figure 7.7 shows the two-dimensional H-field radiation patterns of the proposed antenna in the resonant bands. The H-field radiation pattern is obtained in the azimuth plane (the plane of the antenna). The plots show a

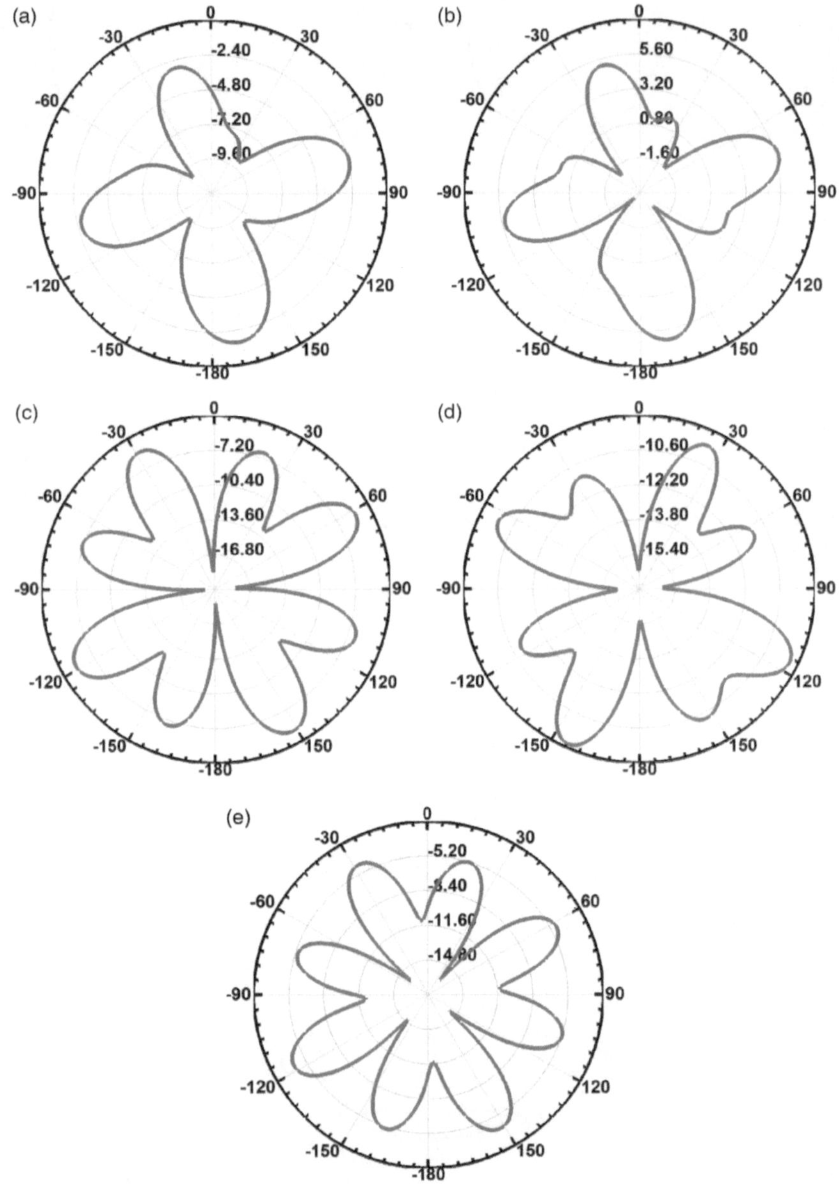

Figure 7.7 Two-dimensional H-field radiation patterns at (a) $f = 3.56$ GHz; (b) $f = 3.94$ GHz; (c) $f = 5.58$ GHz; (d) $f = 6.18$ GHz; and (e) $f = 6.5$ GHz.

full-flourished symmetrical H-field radiation pattern in all the bands of the proposed MIMO antenna. This behavior corresponds to the orthogonal arrangement of the antenna elements.

7.3.5 Three-dimensional radiation results

The three-dimensional radiation plots are shown in Figure 7.8. The plots at $f = 3.56$ GHz and 3.94 GHz have a quad-directional symmetrical radiation pattern radiating in four directions. The plots at $f = 5.58$ GHz and 6.5

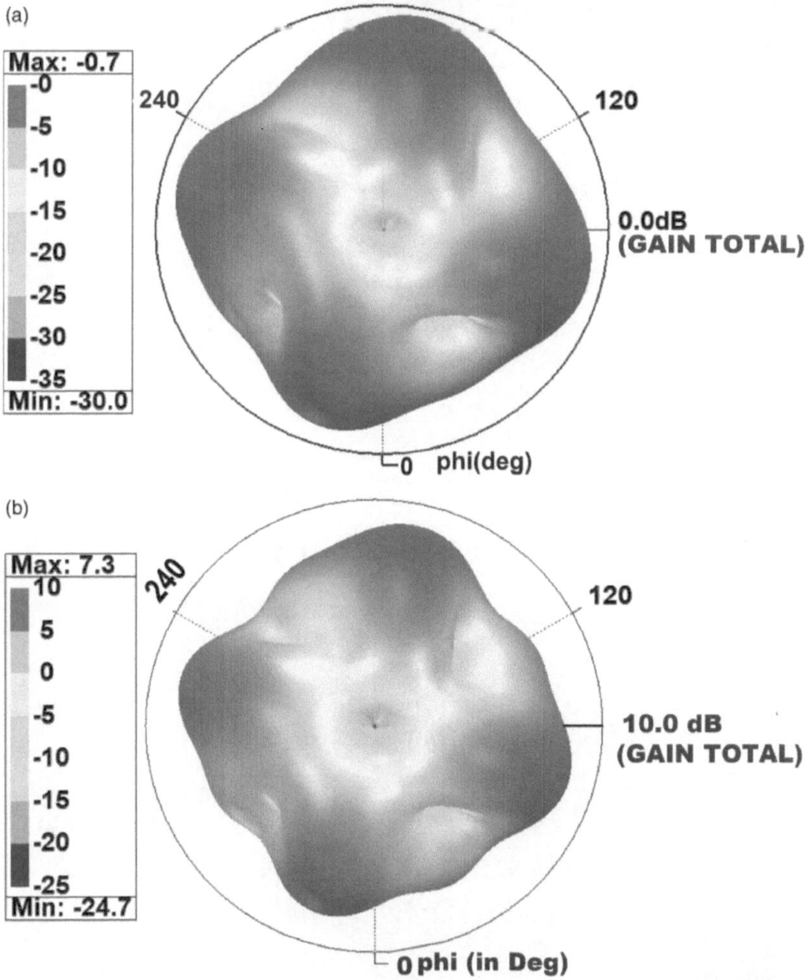

Figure 7.8 Three-dimensional polar plot of proposed MIMO antenna at (a) $f = 3.56$ GHz; (b) $f = 3.94$ GHz; (c) $f = 5.58$ GHz; (d) $f = 6.18$ GHz; and (e) $f = 6.5$ GHz.

(Continued)

(c)

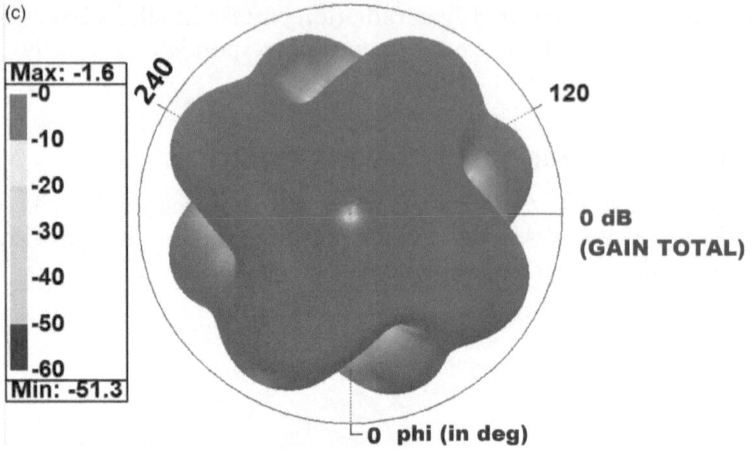

(d)

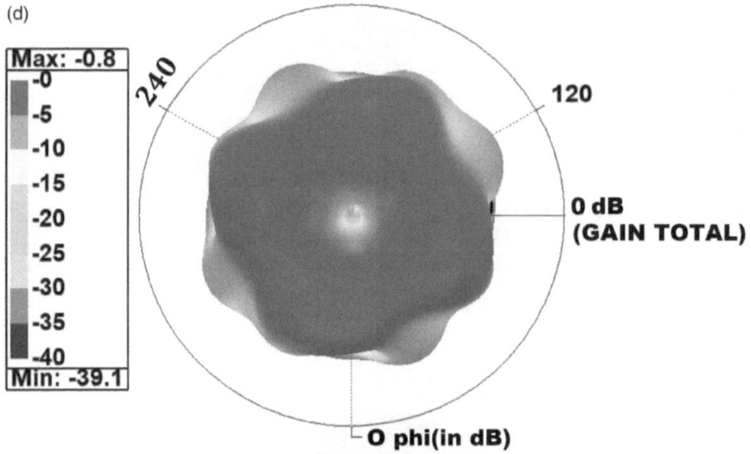

(e)

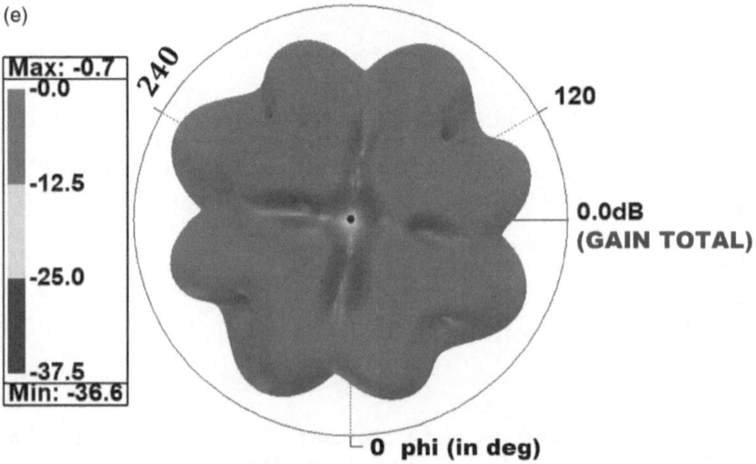

Figure 7.8 (Continued)

Table 7.3 Antenna parameter overlay at the resonating bands

Frequency (GHz)	*P*	*Q*	*R*	*O*	*S*	*T*	*u*	*v*	*w*	*x*	*y*	*z*
3.56	3.45	6.08	0.36	0.34	10.66	39.85	40	0.27	12.79	0	40	0.354
3.94	1.58	5.54	−3.03	−3.04	5.55	39.9	40	0.14	3.38	0	40	−3.04
5.58	2.01	4.67	−1.95	−1.99	8.61	39.65	40	0.22	10.89	0	40	−1.99
6.18	4.4	7.7	1.41	1.41	9.4	39.92	40	0.23	13.49	0	40	1.40
6.5	3.49	8.14	0.42	0.4	6.74	39.8	40	0.17	11.08	0	40	0.4

GHz have a nearly omnidirectional radiation pattern. The plot at $f = 6.18$ GHz has upward and downward radiating fields. The radiation patterns for all the bands of resonance have spatially patterned configurations. This behavior corresponds to the orthogonal configuration of all four antenna elements with respect to each other.

Table 7.3 shows the radiation intensity, directivity, gain, power, efficiency, and front-to-back ratio parameters of the proposed MIMO antenna at the resonating frequencies.

The values for the parameters in Table 7.3 are explained as follows [2]:

P: maximum radiation intensity (U). It signifies the amount of power radiated per unit solid angle. Its unit is mW/sr.

$$U = 10 \times \log \text{power density } (W) \times r^2 \tag{7.10}$$

Q: peak directivity (D). It is the measure of how concentrated the radiation pattern of an antenna is in a specific direction. It is calculated in dB.

$$D = 10 \times \log \frac{4\pi U_{\text{max}}}{P_{\text{accepted}}} \tag{7.11}$$

R: peak gain (G). It is the ratio of the radiation intensity of an antenna in a specific direction to the radiation intensity average over all directions. It is calculated in dB.

$$G = \text{directivity } (D) \times \text{efficiency } (K) \tag{7.12}$$

O: peak realized gain. It is the maximum realized gain over all the user-specified directions of the far-field infinite sphere.

$$\text{Realized gain} = 4\pi \frac{U}{P_{\text{incident}}} \tag{7.13}$$

S: radiated power (P_{rad}) or effective isotropic radiated power (EIRP). It is the amount of power that could be radiated by an antenna. It is calculated in dB.

$$\begin{aligned}\text{EIRP} = {} & \text{output power of transmitter } (P_T) - \text{cable loss } (L_C) \\ & + \text{isotropic antenna gain } (G)\end{aligned} \tag{7.14}$$

T: accepted power (P_{accepted}). It corresponds to the power entering the antenna through one or more ports. It is calculated in mW.
u: incident power (P_{incident}). It corresponds to the input power provided to the antenna. It is calculated in mW.
v: radiation efficiency (e). It calculates the efficiency of the antenna. It is unitless.

$$e = \frac{\text{gain } (G)}{\text{directivity } (D)} \tag{7.15}$$

w: front-to-back ratio. It calculates the ratio of the power in the front/main lobe to the power in the back lobe. It is calculated in dB.

$$\text{Front to back ratio} = \frac{\text{forward power } (F)}{\text{backward power } (B)} \tag{7.16}$$

x: decay power. It is the real part of propagation constant (k).
y: system power (P_{system}). It is the power that is applied by the system simulator. It is calculated in dB.
z: peak system gain (G_{system}). It is the peak gain obtained by the system. It is calculated in dB.

7.3.6 Cross-pole and co-pole radiation patterns

Ludwig (1973) discussed the process for obtaining cross-polarization and co-polarization of an antenna [20]. Co-pole is the desired polarization of antenna radiation, while cross-pole is the undesired polarization of antenna radiation. Figure 7.9 shows the plots at the resonant bands.

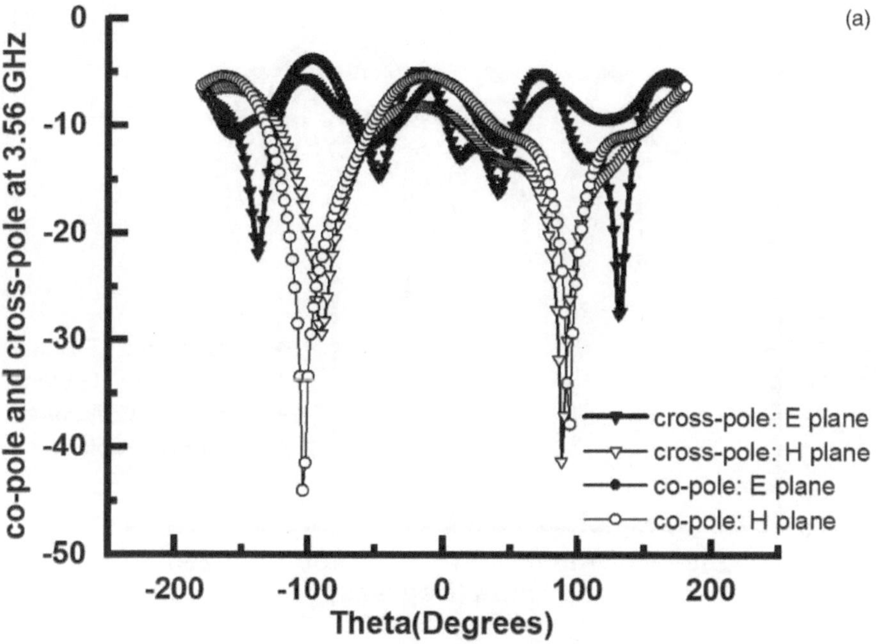

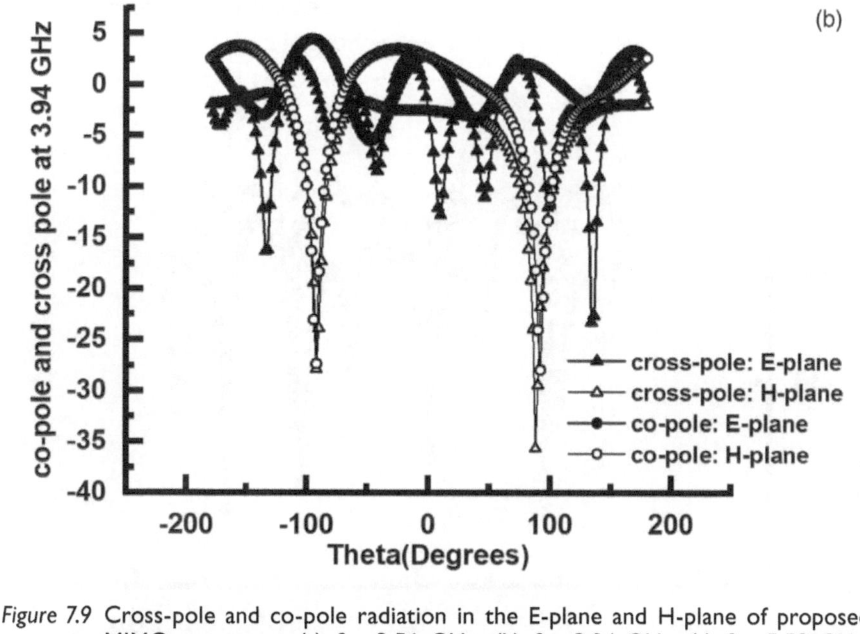

Figure 7.9 Cross-pole and co-pole radiation in the E-plane and H-plane of proposed MIMO antenna at (a) *f* = 3.56 GHz; (b) *f* = 3.94 GHz; (c) *f* = 5.58 GHz; (d) *f* = 6.18 GHz; and (e) *f* = 6.5 GHz.

(Continued)

(c)

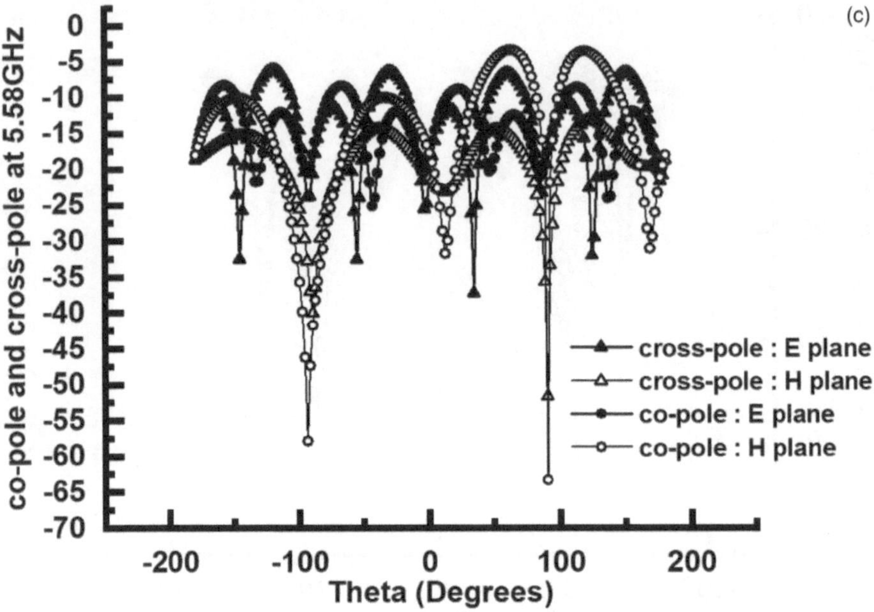

(d)

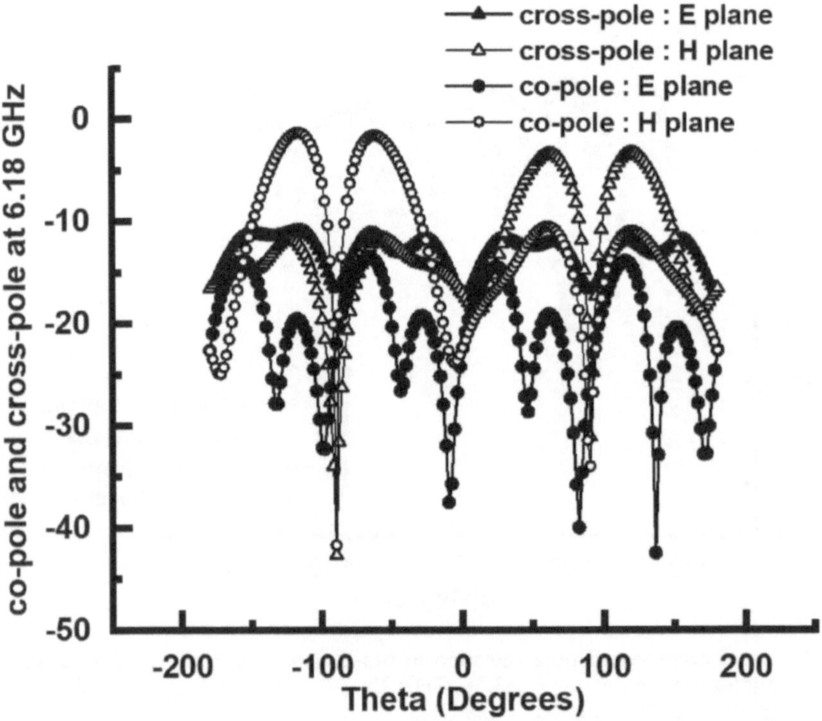

Figure 7.9 (Continued)

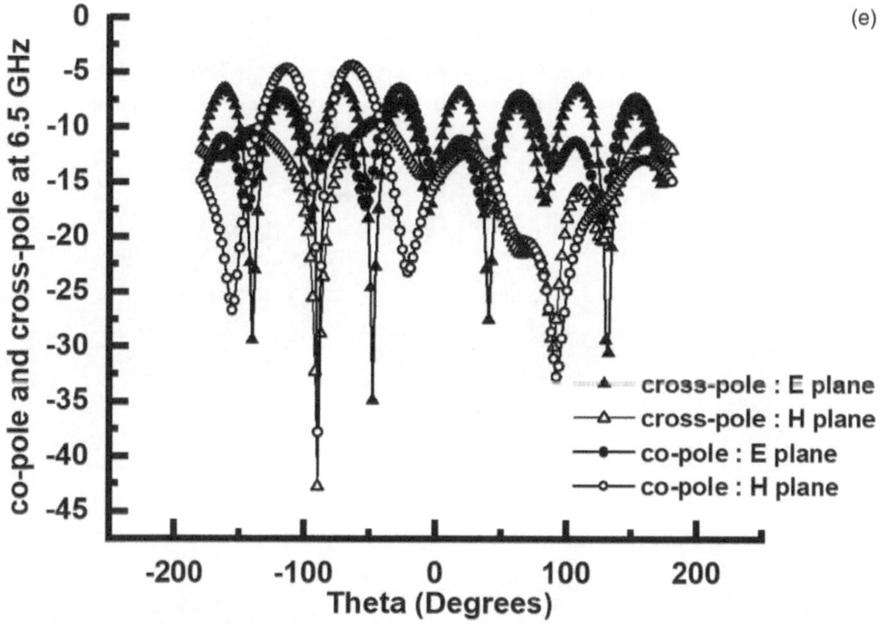

Figure 7.9 (Continued)

7.3.7 Current distribution pattern

Figure 7.10 shows the surface current distribution pattern of the proposed antenna at f = 3.56, 3.94, 5.58, 6.18, and 6.5 GHz. The current distribution pattern, as shown in Figure 7.10, has a patterned distribution of least current in most of its regions for all the bands. An entire least current distribution for the antenna can be seen at f = 5.58 GHz. Also, the antenna has maximum surface coverage of least current distribution at f = 3.56, 6.5, and 6.18 GHz. For f = 3.94 GHz, some areas are deprived of least surface current distribution.

7.4 MIMO PARAMETERS

7.4.1 Isolation plots

The transmission coefficient gives the value of the power transmitted from one port to another in an antenna operation. Figure 7.11 gives the isolation plots of four antennas with respect to each other.

For the proposed MIMO antenna, $S_{21} = S_{12}$, $S_{13} = S_{31}$, $S_{14} = S_{41}$, $S_{23} = S_{32}$, $S_{24} = S_{42}$, and $S_{34} = S_{43}$. Hence, one from each pair is used to plot the

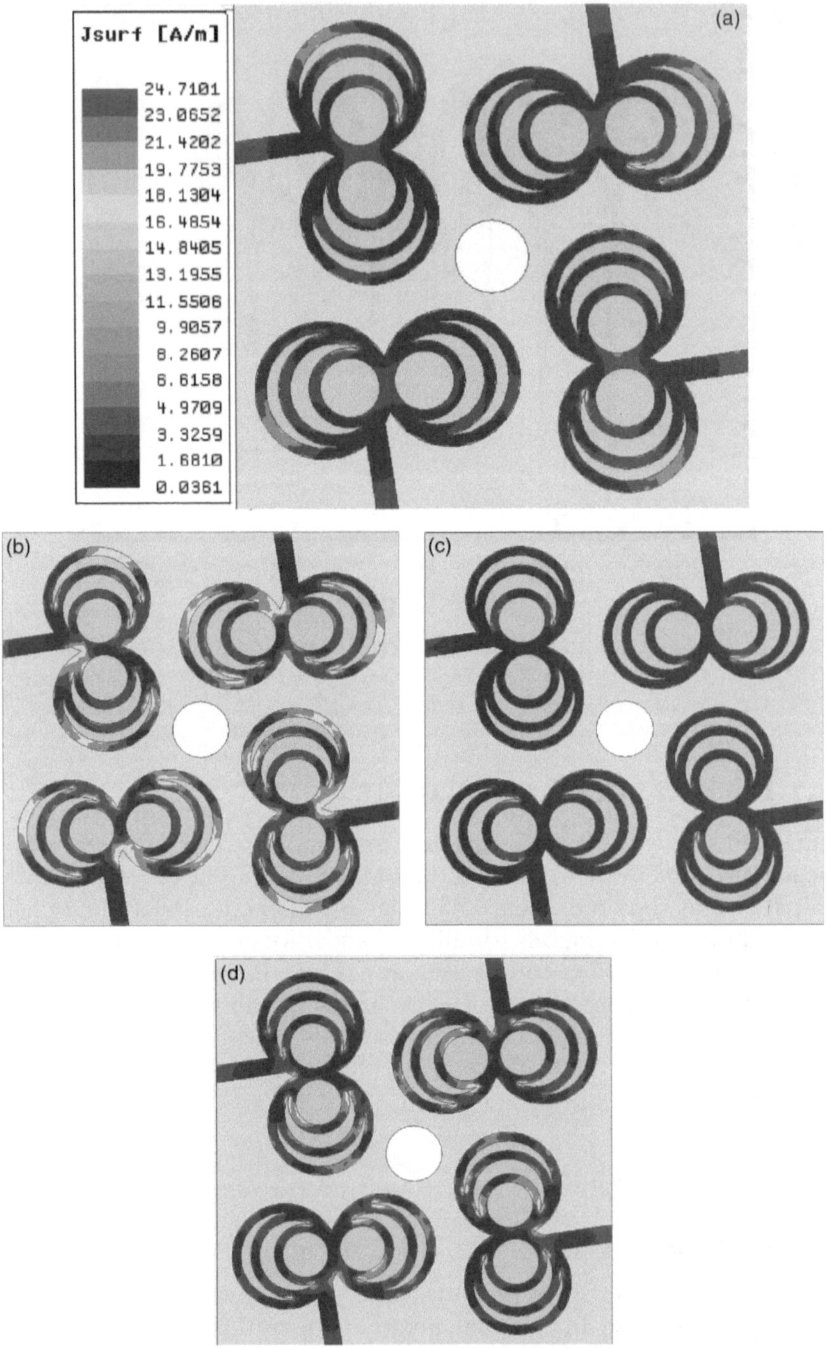

Figure 7.10 Surface current distribution patterns at (a) $f = 3.56$ GHz; (b) $f = 3.94$ GHz; (c) $f = 5.58$ GHz; (d) $f = 6.18$ GHz; and (e) $f = 6.5$ GHz.

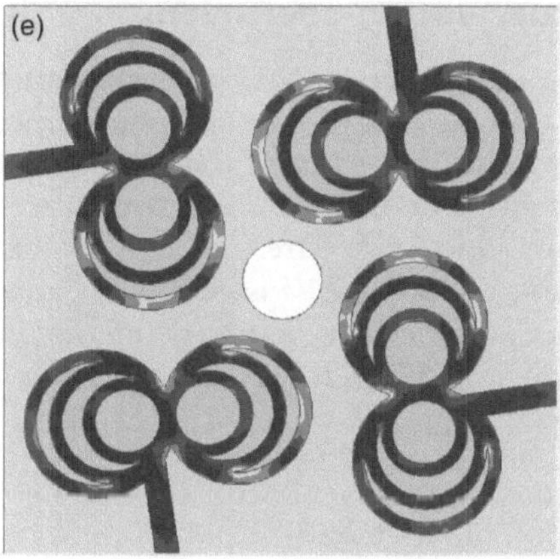

Figure 7.10 (*Continued*)

isolation graph. The difference between the transmission coefficient and the reflection coefficient gives the isolation of the transmitted and radiated powers. More isolation is desirable for the efficient working of the antenna. The proposed design shows an isolation of 4 to 5 dB in all the bands of operation. The value of isolation is denoted as S_{21} or the transmission coefficient of the antenna.

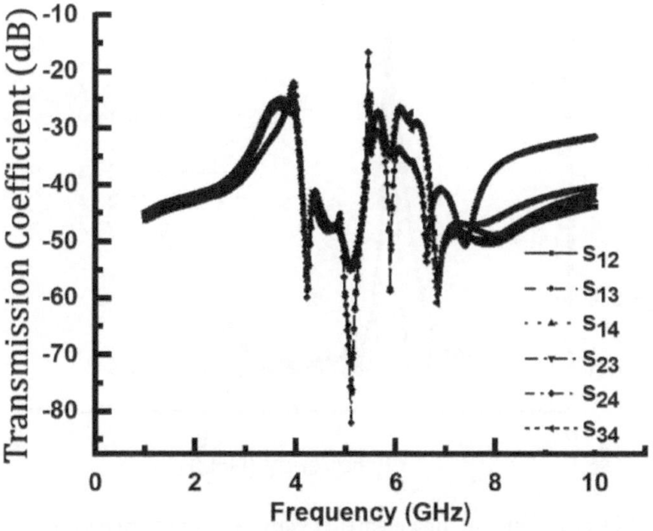

Figure 7.11 Transmission coefficient (S_{21}) plots of the proposed MIMO antenna.

7.4.2 Envelope correlation coefficient

The ECC value is an important MIMO antenna parameter. It calculates how much the radiation pattern of one antenna element differs from the patterns of the others. The ECC value of the antenna is calculated in terms of S parameters or is based on far-field radiation pattern values. However, in high-frequency simulation software, the ECC values are calculated in terms of S parameters. Equation 7.17 is the ECC equation of an antenna [21]. The plots of ECC and diversity gain are obtained based on the logistics presented by Ludwig (1973) [20]. Figure 7.12 shows the ECC plot of the proposed antenna.

$$\rho_{ECC} = -\frac{S_{11}S_{12}^{*} + S_{21}S_{22}^{*}}{\sqrt{\left(1-|S_{11}|^2-|S_{21}|^2\right)\left(1-|S_{22}|^2-|S_{12}|^2\right)\eta_1\eta_2}} \tag{7.17}$$

The ECC values of the antenna proposed in this chapter are in good agreement, which is desired. The values lie below 0.2 for the entire range except in the 3.82 to 4.06 GHz band, where the value of the ECC lies between 0.5 and 1. The antenna resonates at 3.4 to 3.98 GHz, but as the value of the ECC lies in an unfavorable range, using the antenna for this range proves worthless. Enhancements in the design can be employed to provide a good ECC value in this range.

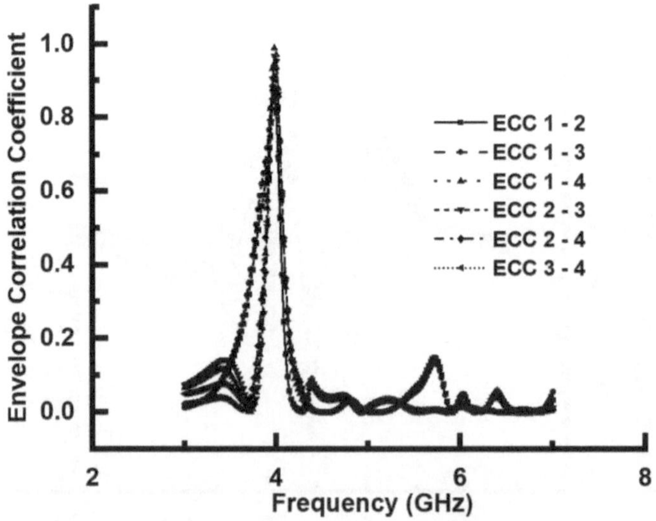

Figure 7.12 ECCs of all four antenna elements with respect to each other.

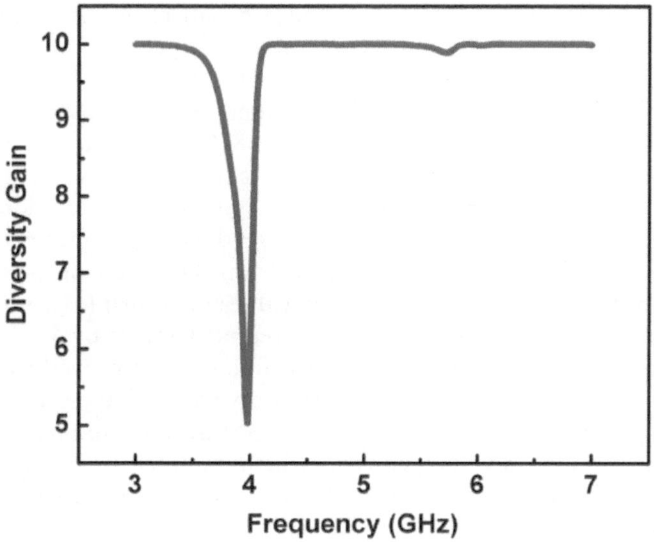

Figure 7.13 Diversity gain of proposed MIMO antenna.

7.4.3 Diversity gain

For a value of the bit error rate (BER), the diversity gain parameter of a MIMO antenna measures the decrease required to receive the SNR value averaged over the fading. The diminished fading is obtained by reducing the fading of the smart antenna. The diversity gain is obtained using ECC values, as shown in Equation 7.18. Its value should be near 10 dB in the operating range of frequencies, corresponding to good MIMO antenna parameter values [22]. The plot of diversity gain is shown in Figure 7.13.

$$DG = 10 \times \sqrt{1 - |ECC|} \tag{7.18}$$

The values are in good agreement in the entire range except in the region corresponding to the 3.82 to 4.06 GHz band. There is a sharp dip in the plot obtained in this region of operation of the antenna.

7.5 CONCLUSION

The antenna proposed in this chapter resonates in three frequency bands between 3 to 7 GHz. These frequency bands are 3.42 to 3.98 GHz, 5.52 to 5.66 GHz, and 6.04 to 6.58 GHz. The peak resonant frequencies were found to be 3.56, 3.94, 5.58, 6.18, and 6.5 GHz. The bandwidths obtained are 580, 140, and 540 MHz, providing a broad range of coverage. The ECC values lies below 0.2 for the entire range of operation, meeting industrial specifications (that is, ECC < 0.5) except in the 3.82–4.04 GHz band, where the ECC is greater than 1. The antenna covers the pioneer fifth generation (5G) band corresponding to 3.4–3.8 GHz (auctioned 5G band). Also, the ECC value obtained in this region is in the valuable range. Hence, the antenna is suitable for 5G applications. Symmetrical radiation patterns are obtained for each band of resonance. To reduce the surface wave excitation, a small circular portion is removed from the center of the substrate. It showed a considerable increase in the values of return loss, VSWR, and other radiation parameters. Tilting the antenna elements by 5.67° from the horizontal and tilting the microstrip feed line by 8.5° proved valuable for achieving angle, spatial, and polarization diversity. The design presented has an outgoing wave-like structure for the patch. A ground plane designed for radiation proves advantageous for the symmetricity in the radiation pattern. The design can be further enhanced, keeping in view the radiation potential of the antenna.

REFERENCES

[1] M. S. Sharawi, *Printed MIMO Antenna Engineering* (Norwood, MA: Artech House, 2013).

[2] C. A. Balanis, *Antenna Theory: Analysis and Design*, 4th ed. (Hoboken, NJ: Wiley, 2016).

[3] S. W. Su, "High-gain dual-loop antennas for MIMO access points in the 2.4/5.2/5.8 GHz bands," *IEEE Transactions on Antennas and Propagation*, vol. 58, no. 7, pp. 2412–2419, July 2010, doi: 10.1109/TAP.2010.2048871

[4] M. Han and J. Choi, "Multiband MIMO antenna using orthogonally polarized dipole elements for mobile communications," *Microwave and Optical Technology Letters*, vol. 53, no. 9, pp. 2043–2048, Sep. 2011, doi: 10.1002/mop.26198

[5] M. S. Han and J. Choi, "Multiband MIMO antenna with independent resonance frequency adjustability," *Microwave and Optical Technology Letters*, vol. 52, no. 8, pp. 1893–1901, Aug. 2010, doi: 10.1002/mop.25334

[6] L. Malviya, R. K. Panigrahi, and M. V. Kartikeyan, "Four element planar MIMO antenna design for Long-Term Evolution operation," *IETE Journal of Research*, vol. 64, no. 3, pp. 367–373, 2018, doi: 10.1080/03772063.2017.1355755

[7] P. Kaur, G. Singh, J. Singh, and M. Kaur, "Design and development of symmetrical E-shaped microstrip patch antenna for multiband wireless applications," *International Journal of Signal Processing, Image Processing and Pattern Recognition*, vol. 9, no. 11, pp. 405–416, 2016, doi: 10.14257/ijsip.2016.9.10.38

[8] A. Kaur, G. Singh, and M. Kaur, "Miniaturized multiband slotted microstrip antenna for wireless applications," *Wireless Personal Communications*, vol. 96, no. 1, pp. 441–453, 2017, doi: 10.1007/s11277-017-4177-4
[9] M. S. Sharawi, A. B. Numan, and D. N. Aloi, "Isolation improvement in a dual-band dual-element MIMO antenna system using capacitively loaded loops," Progress in Electromagnetics Research, vol. 134, pp. 347–266, 2013.
[10] S. Zhang, B. K. Lau, A. Sunesson, and S. He, "Closely-packed UWB MIMO/ diversity antenna with different patterns and polarizations for USB dongle applications," *IEEE Transactions on Antennas and Propagation*, vol. 60, no. 9, pp. 4372–4380, 2012, doi: 10.1109/TAP.2012.2207049
[11] X. Zhou, X. Quan, and R. Li, "A dual-broadband MIMO antenna system for GSM/UMTS/LTE and WLAN handsets," *IEEE Antennas and Wireless Propagation Letters*, vol. 11, pp. 551–554, 2012, doi: 10.1109/ LAWP.2012.2199459
[12] A. N. Kulkarni and S. K. Sharma, "Frequency reconfigurable microstrip loop antenna covering LTE bands with MIMO implementation and wideband microstrip slot antenna all for portable wireless DTV media player," *IEEE Transactions on Antennas and Propagation*, vol. 61, no. 2, pp. 964–968, 2013, doi: 10.1109/TAP.2012.2223433
[13] C. Yang, J. Kim, H. Kim, J. Wee, B. Kim, and C. Jung, "Quad-band antenna with high isolation MIMO and broadband SCS for broadcasting and telecommunication services," *IEEE Antennas and Wireless Propagation Letters*, vol. 9, pp. 584–587, 2010, doi: 10.1109/LAWP.2010.2053515
[14] S. C. Fernandez and S. K. Sharma, "Multiband printed meandered loop antennas with MIMO implementations for wireless routers," *IEEE Antennas and Wireless Propagation Letters*, vol. 12, pp. 96–99, 2013, doi: 10.1109/ LAWP.2013.2243104
[15] A. Foudazi, H. R. Hassani, and S. M. A. Nezhad, "A dual-band WLAN/ UWB printed wide slot antenna," in 2011 Loughborough Antennas and Propagation Conference, 2011, pp. 1–3, doi: 10.1109/LAPC.2011.6114065
[16] M. Han and J. Choi, "Dual-band MIMO antenna using a symmetric slotted structure for 4G USB dongle application," in IEEE Antennas and Propagation Society, AP-S International Symposium (Digest), 2011, pp. 2223–2226, doi: 10.1109/APS.2011.5996957
[17] Y. Cheon, J. Lee, and J. Lee, "Quad-band monopole antenna including LTE 700 MHz with magneto-dielectric material," IEEE Antennas and Wireless Propagation Letters, vol. 11, pp. 137–140, 2012, doi: 10.1109/ LAWP.2012.2184517
[18] S. Hong, J. Lee, and J. Choi, "Design of UWB diversity antenna for PDA applications," in 2008 10th International Conference on Advanced Communication Technology, 2008, pp. 583–585.
[19] Y. Kumar and S. Singh, "Performance analysis of coaxial probe fed modified Sierpinski–meander hybrid fractal heptaband antenna for future wireless communication networks," *Wireless Personal Communications*, vol. 94, no. 4, pp. 3251–3263, 2017, doi: 10.1007/s11277-016-3775-x
[20] A. C. Ludwig, "The definition of cross polarization," *IEEE Transactions on Antennas and Propagation*, vol. 21, no. 1, pp. 116–119, 1973, doi: 10.1109/ TAP.1973.1140406

[21] R. Cornelius, A. Narbudowicz, M. J. Ammann, and D. Heberling, "Calculating the envelope correlation coefficient directly from spherical modes spectrum," 2017 11th European Conference on Antennas and Propagation, 2017, pp. 3003–3006, doi: 10.23919/EuCAP.2017.7928132

[22] A. W. Mohammad Saadh, K. Ashwath, P. Ramaswamy, T. Ali, and J. Anguera, "A uniquely shaped MIMO antenna on FR4 material to enhance isolation and bandwidth for wireless applications," *AEU—International Journal of Electronics and Communications*, vol. 123, p. 153316, 2020, doi: 10.1016/j.aeue.2020.153316

Chapter 8

5G communication challenges and opportunities

Shaping the future

Shalini Sah and Devica Verma

CONTENTS

DOI: 10.1201/9781003290230-8

8.1 INTRODUCTION

The progression of wireless communication technologies over the past two decades started with the introduction of cellular 2G Global System for Mobile (GSM) and reached advanced Long-Term Evolution (LTE); today is the era of 5G. New technologies keep coming up as researchers improve over the older ones and provide better throughput, low latency, more bandwidth, etc. Throughput is defined as the rate at which data is transferred, whereas the processing speed of each node through which the data streams are traversing gives the latency [1].

As the technology evolved from 2G to the 3G Universal Mobile Telecommunication System (UMTS), real-time video calls became possible due to faster system and download speeds. There were other benefits like increased network capacity, which is the basic purpose of cellular communication, and also reduced delay in accessing application servers and providing data, voice, and video access remotely at any time and place. All this was made possible by LTE and the resulting -Advanced (LTE-A) [1]. The first digital mobile voice communication standards were provided by 2G, which had better coverage, whereas 3G was initially planned for voice with some mixed media consideration. The information rate improved as the technologies evolved one after the other: 2G speed is 64 kbps, 3G is 2 Mbps, and 4G is 50–100 Mbps. Today, companies are working hard to upgrade the data transfer speed through 5G, which will help in the improvement of other network parameters like connectivity and scalability.

When 5G networks become a reality, the operation of all devices and critical appliances can be controlled over the networks with almost zero delay. The internet of things (IoT) can be made accessible to all, and. hence, the idea of continuously monitoring machines through mobiles will be conceivable. Based on the requirements of the user and the various use cases, a single network can be used to provide different services.

8.2 EVOLUTION OF MOBILE TECHNOLOGIES

Mobile technologies are evolving and showing the trend in communication, which initially started with 1G analog voice and has grown to 5G, embedding the latest technologies like artificial intelligence (AI) and cloud computing (see Figure 8.1).

First Generation (1G): The 1980s saw the emergence of 1G technology, which consisted of an analog system and gave birth to cell phones. Many mobile technologies were introduced during this time like the Advanced Mobile Phone System (AMPS) developed by Bell Labs, which was an improved version of Improved Mobile Telephone Service. 1G was analog voice with a frequency range of 150 MHz. The multiple

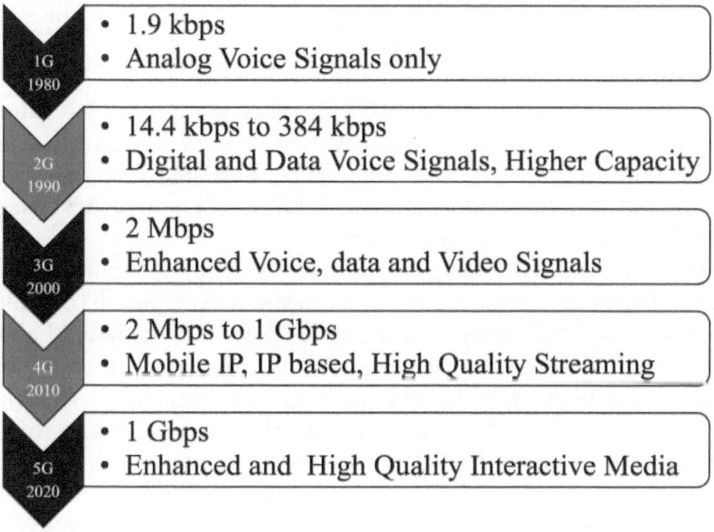

Figure 8.1 Evolution of mobile technologies.

access technology used in 1G is the frequency division multiple access (FDMA) technique. The capacity and coverage of 1G were very limited, and the technology also suffered from unpredictable handoff, feeble voice quality, and vulnerability to undesirable eavesdropping [2].

Second Generation (2G): 2G simply refers to the mobile technology that was based on digital technology and deployed in the early 1990s in Finland. It used bandwidth from 30 to 200 KHz. It provided services like text messages, picture messages, and Multimedia Messaging Service. This technology offered greater security for the sender as well as the receiver compared to IG technology. 2G used both time-division multiple access (TDMA) and code-division multiple access (CDMA). TDMA utilizes different time slots for sending the signal, and CDMA utilizes different codes for the same. GSM technology was used in 2G. Then came the development of the 2.5G system, which utilized both circuit switching and packet switching. Data rates were up to 144 kbps in 2G deployment.

Third Generation (3G): 3G incorporated data services, access to television/video, verbal communication, and new services like Global Roaming. For clarity enhancement, the wideband wireless network was used in 3G. Technology called packet switching was implemented for data transfer, and circuit switching technology was utilized for the interpretation of voice calls. 3G operated at a range of 2100 MHz and a bandwidth of 15–20 MHz, which made video calling services and high-speed internet a reality. The use of wideband voice channels by 3G converted the whole world into a small village, as it enabled people to contact each other and even send messages while sitting in two different corners of the world [2].

Fourth Generation (4G): 4G refers to the widespread wireless networks that replaced the 3G networks. 4G offers a much wider bandwidth than 3G. In 3G, the bandwidth was 5 MHz, while in 4G, it ranged from more than 20 MHz to 100 MHz. The data rate also improved significantly. The date rate was 1 Gbps for stationary users and 100 Mbps for high-mobility users.

Fifth Generation (5G): 5G mobile technology is better than the previous technologies in several ways. 5G occupies a very high bandwidth and thus is capable of changing the user experience like never before. 5G offers high data speeds up to 20 Gbps, very low latency, high reliability, and high capacity and coverage. The improved efficiency and performance of the 5G networks make them a good choice to connect new industries and enhance the user experience. 5G is going to impact almost every industrial sector, be it transportation, health care, or manufacturing. 5G is capable of providing a greater capacity, since it will expand into a new spectrum called millimeter wave (mmWave).

A comparison of the generations of mobile technology is shown in Table 8.1. This table includes the frequency band, data bandwidth, and basic requirements that are an essential part of the technology.

8.2.1 Why is 5G required?

5G is the key to the growth of the internet and various devices connected to it and will empower various industries to connect over the internet. 5G incorporates other features [4]:

- Better capacity and coverage
- Lower battery power consumption
- Data transfer paths that are multiple and concurrent
- Mobility rate of around 1 Gbps
- Enhanced security
- Spectral efficiency of higher system level
- Worldwide wireless web
- AI is the core of all applications
- Low infrastructure deployment costs that facilitate cheaper traffic fees [5]

8.2.2 Characteristics of 5G technology

- 5G uses high resolution to provide fast internet access to consumers.
- The most convenient and beneficial feature of 5G technology is that it provides billing limits beforehand.

Table 8.1 Comparison of various mobile technologies [3]

Technology	*1G*	*2G*	*3G*	*4G*	*5G*
Basic requirements	Analog voice technology	Digital technology, data rate is 64 kbps	Worldwide mobile broadband, data rate is 2 Mbps	Advanced IMT, data rate is 2 Mbps	High data rates and low latency to support cloud services and mobile services
Bandwidth	1.9 kbps	14.4–384 kbps	2 Mbps	2 Mbps–1 Gbps	More than 1 Gbps as per demand
Services offered	Analog voice using human- human voice communication	Digital voice, digital data communication	Audio, video, data, high-definition quality	High-definition streaming, dynamic information access, global roaming, wearable devices, wearable devices roaming	High-definition streaming, dynamic information access, wearable devices with reduced latency
Standards	NMT, AMPS, Hicap, CDPD, TACS, ETACS	GSM, GPRS, EDGE, etc.	WCDMA, CDMA 2000	Al access includes OFMDA, MC-CDMA, Network-LMPS	OFDMA, BDMA
Multiple access techniques	FDMA	TDMA, CDMA	CDMA	CDMA	OFDMA, BDMA
Year of implementation	1970–1984	1990	2001	2010	2015
Switching techniques	Circuit switched	Circuit and packet switched	Circuit and packet switched	Packet and message switched	Full packet switched
Frequency band	800–900 MHz	850–1900 MHz	1.6–2.5 GHz	2–8 GHz	600 MHz–71 GHz

- 5G allows its mobile users the option of printing their telephone records.
- It allows data in gigabits.
- 5G carrier distribution gateways have unmatched maximum stability.
- Results obtained from the information transfer technology are more accurate and reliable while using 5G.
- 5G provides the same comfort to consumers, even if they are using the technology remotely, due to increased speed, low latency, and high reliability.
- Virtual private networks are supported by 5G.
- 5G networks have very high uplink and downlink speeds.
- They provide enhanced and advanced connectivity to almost every corner of the planet.
- Speed and reliability are the hallmarks of 5G networks.

8.2.3 Applications of 5G technology

- Helpful in creating a real wireless world where there are no limitations imposed by accessibility and time zone.
- AI-enabled wearable devices.
- Development of Internet Protocol version 6 in which a visiting care-of mobile IP address is made consistent with location as well as connected network.
- User has the advantage of simultaneously getting connected with many wireless access technologies such as 2.5G, 3G, 4G, 5G, Wireless Fidelity (Wi-Fi), and wireless personal area network (WPAN) or the future access technologies and can smoothly move between them. 5G is a step ahead with the concept of multiple data transfer paths.
- Cognitive radio technology finds unused spectrum and distributes it equally among different radio technologies and adapts the transmission scheme according to the technologies that are presently sharing the spectrum. Dynamic radio spectrum management is obtained through division and allocation and depends on software-based radio, also known Smart Radio.
- A Korean research and development program that supported beam division multiple access and group cooperative relay techniques [4] recommended the use of the radio interface of 5G communication systems.

8.3 THE BACKDROP OF 5G

Cellular network technology has been progressing since the 1980s [2]. 5G will enhance the mobile network, which, in turn, will allow for more connections. This connectivity offered by 5G will help to unleash a significant amount of potential for various industries.

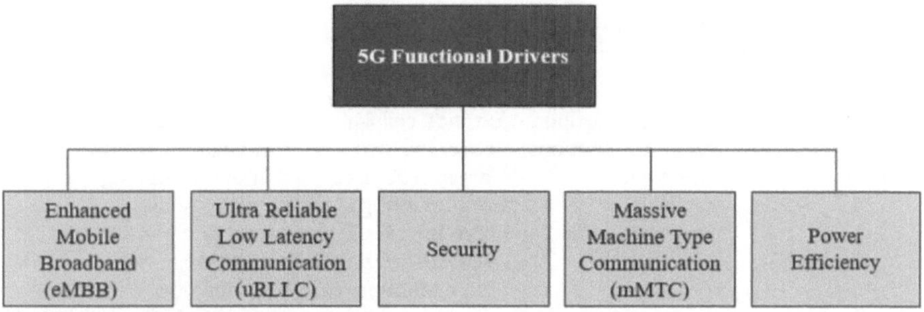

Figure 8.2 Functional drivers of 5G.

The transition from the earlier generations to 5G presents several new features that make it unique and more advanced than previous generations. 5G will create new value in our society and industries. Figure 8.2 depicts the five functional drivers of 5G. These key drivers [6] along with their use cases are described in Table 8.2.

8.3.1 Spectrum allocated to different technologies

In simple terms, wireless signals travel over a portion of the electromagnetic spectrum assigned to the mobile/cellular industry (20 KHz–300 GHz).

- AM and FM channels: 100–200 MHz.
- Telecom spectrum: 800–2300 MHz.
- Mobile communication: 800, 900, 1800, 2100, and 2300 MHz.
- 2G services were launched in India at 900 MHz and later 1800 MHz, which is also getting used for 4G.
- 3G mobile technology is available at both 900 and 2100 MHz in India.
- Beyond 2300 MHz are unlicensed bands used for Wi-Fi and Bluetooth; Wi-Fi used to be 2.4 GHz (2400 MHz) and has begun to shift to the 5 GHz band.

8.3.2 Spectrum in the case of 5G

In wireless communication, spectrum can be divided into three categories: low-, mid-, and high-band spectrum shown in figure 8.3.

In case of 5G networks, we need all the three bands:

- *Low-band spectrum:* It has minimal signal interruptions while traveling longer distances.
- *High-band spectrum:* Its traveled distances are shorter with ultra-fast speeds and high capacity.
- *Mid-band spectrum:* It is a mix of the characteristics of both low- and high-band spectrum.

Table 8.2 Functional drivers of 5G

Key drivers	*Description*	*Values*	*Use cases*
Enhanced mobile broadband (eMBB)	Higher throughput, faster connections, and greater capacity	Extends cellular coverage into large structures Handles a huge number of devices utilizing large amounts of data	Remote surgery and examination, massive content streaming services, public protection and disaster response services, high-definition cloud gaming, enhanced digital signage, dense area service, real-time AR/VR, wired and wireless access service, improved indoor broadband service
Ultra-reliable low latency communication (uRLLC)	Reduces the time taken to upload data from a device to the target	Time-sensitive connections where we cannot afford to have a delay are enabled wirelessly	Mission critical services, HD real-time gaming, autonomous vehicles, drones and robotic applications, health monitoring systems, intelligent transportation, factory automation, remote operation, smart grid and metering, self-driving cars
Security	Offers robust security properties, which slow high availability and reliability	Supports applications where failure is not an option by creating an ultra-reliable connection	
Massive machine-type communication (mMTC)	Small cell deployment and increased spectral efficiency	Offers a huge number of connections to support data-intensive applications	Smart cities/buildings/agriculture, internet of energy/utility management, industrial automation, asset tracking and predictive maintenance, smart logistics, smart consumer variables, environmental management, intelligent surveillance and video analytics, smart retail
Power efficiency	Efficient power requirement for MIMO, small cell implementation	Allows massive IoT and leads to lower costs	

Source: ITU, 2018 [6].

8.3.3 Access network

The telecommunication network that connects the subscribers to their service providers is referred to as the access network. It may also refer to the series of cables, wires, and equipment through which a telephone link reaches the telephone user and the local exchange.

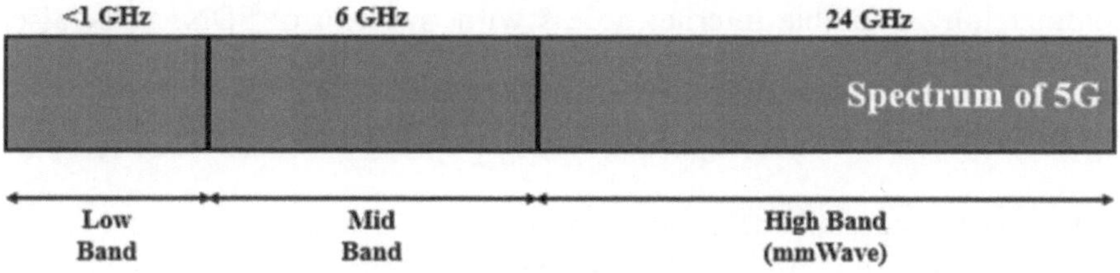

Figure 8.3 Band spectrum for 5G: low, high, and middle.

A radio access network (RAN) is basically a part of mobile telecommunication that connects devices to the other part of network via radio connections. It is located between a cellular phone, a personal computer, or any remotely controlled machine and its core network, and its functionality is usually controlled by chips inside both the core network and the consumer device.

The core network is a part of a telecommunication network. The main functions of the core network are to offer different services to the end user and to connect direct telephone calls over the public-switched telephone network (PSTN).

8.3.4 Network functions virtualization and software-defined networking

Network abstraction is used by both network functions virtualization (NFV) and software-defined networking (SDN). Decoupling the network functions from the proprietary hardware and then running them as software in virtual machines (VMs) comprise NFV. Functions like the firewalls, control, and virtual routing are known as the virtual network functions. Both NFV and SDN depend on virtualization to abstract the network infrastructure and design in the software and then to implement the basic software across the hardware devices. Some features of SDN are as follows:

- Makes networking and internet protocol routing flexible
- Decouples control and data plane
- Offloads brain to centralized controller
- Central view of resources
- Programmable network is centrally managed

8.3.5 Software-defined wide area network

A software-defined wide area network (SD-WAN) is employed to simplify the working of a standard wide area network (WAN). The task of traffic management is removed from the physical devices and transferred to software, which provides benefits like increased flexibility and agility. A higher-performance WAN can be created at a much lower cost and provide commercially available internet access with the help of SDN. SDN also allows undertakings to support any of the transport services that securely connect users to applications.

SD-WAN provides certain benefits for enterprises:

- User satisfaction and business productivity are increased.
- Business agility and responsiveness are enhanced.
- Threats are reduced, and security is increased.
- The traditional WAN architecture is simplified.
- WAN costs are reduced by up to 90 percent.

The differences between the traditional WAN network and SD-WAN are described in Table 8.3.

8.3.6 Citizens Broadband Radio Service

The Citizens Broadband Radio Service (CBRS) is a time division duplex (TDD) band that only LTE is allowed to use. The U.S. Federal Communications Commission established the CBRS for wireless broadband sharing in the 3550–3700 MHz band, which is referred to as the 3.5 GHz band. Access to the CBRS will allow commercial property managers to deploy their own private LTE networks with a mixture shared

Table 8.3 Differences between WAN and SD-WAN

WAN	*SD-WAN*
• Expensive bandwidth Limited bandwidth of expensive private/MPLS circuits inhibits rollout and impacts performance of applications	• Virtual WAN Delivers a network overlay with multiple links from different service providers to form a unified pool of bandwidth; offers superior performance and high availability for applications independent of the underlying transport links
• Data center dependence Cloud resources don't have direct access	• Cloud optimized Enterprise-grade performance is assured even when traffic is going to cloud applications
• Complex infrastructure Multitude of single-function appliances are included in traditional WAN, connecting via different WAN links	• Pay-as-you-grow model Since it is on the cloud, it allows the pay-as-you-grow subscription model for cost efficiency
• Less security	• More security

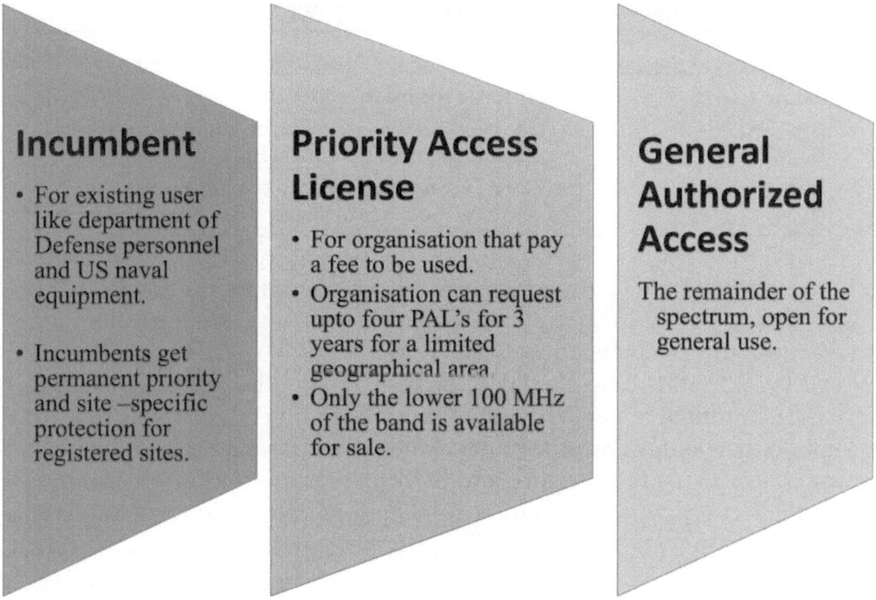

Figure 8.4 Tiers of CBRS spectrum.

unlicensed and licensed parts of the spectrum. With the CBRS, a building owner can give tenants or customers LTE coverage for their devices without paying traditional carriers to do so. When the 3.5 GHz band is opened, more devices are going to be ready to come online with better coverage and capacity. The three tiers of CBRS spectrum use are shown in Figure 8.4.

8.3.7 Private LTE network

A private LTE network is a reliable dedicated network with better coverage for its users, several industries, businesses and IoT; it uses shared licensed and unlicensed parts of the spectrum. Private LTE networks can be based in the U.S. and have the option to operate at either the unlicensed spectrum at 5 GHz or the CBRS band at 3.5 GHz. A private LTE network provides enhanced coverage, capacity, and control.

The Wireless Innovation Forum, the MulteFire Alliance (Qualcomm, Huawei, Boingo, CableLabs, Ericsson, Softbank, Nokia, Samsung, Intel, and Comcast) and the CBRS Alliance are the three industry groups that support private LTE networks in the U.S.

8.3.8 Slicing the 5G network

In simple terms, the division or separation of numerous virtual networks that tend to operate on the same hardware to serve different purposes is termed

Table 8.4 Applications of slicing on the frequency range

Frequency range	*Applications*
Below 1 GHz	Longer range for massive IoT
1–6 GHz	A wider bandwidth is provided for improved mobile broadband and mission-critical applications
Above 6 GHz	mmWave: Extreme bandwidths, shorter range for extreme mobile broadband

network slicing. In 5G networks, network slicing divides the resources into slices (virtual or logical networks). This helps to offer distinct characteristics to the use cases. Each use case requires some specific characteristic offered by 5G, and network slicing provides that. The network operators have the advantage of deploying the functions that are completely essential to support particular customers, particular market segments, and particular use cases. Thus, the network and its resources are optimized. Table 8.4 shows the frequency ranges and the applications offered in those ranges.

Following are some of the benefits offered by network slicing as shown in figure 8.5:

- Allows various services based on the needs of the user and use cases to be offered using a single network.
- Helps network operators to allocate the correct amount of resources required per slice of the network.
- Makes the network and resources efficient and effective: for example, one slice of the network can be used to offer a low data rate and low latency, while the other can be used to offer high throughput.
- Reduces operational expenditure (OPEX) and capital expenditure (CAPEX).
- Improves the time to market for the delivery of 5G network services.

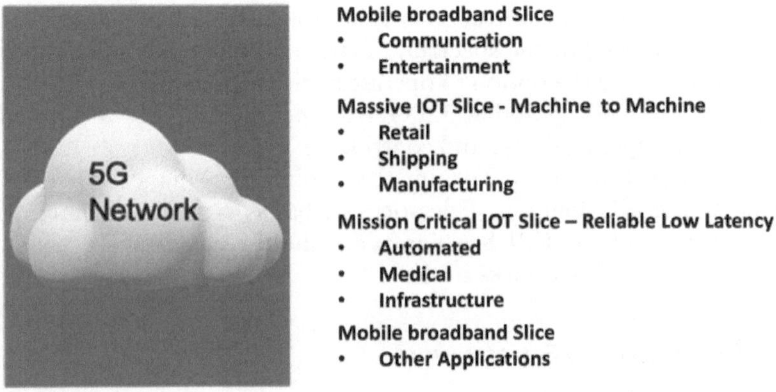

Figure 8.5 Slicing in 5G network.

8.4 INDUSTRIAL AND SOCIAL IMPACTS OF 5G

5G is going to add to industrial advances in three ways:

- Empowering quicker and more effective checking by using predictive intelligence.
- Improving working environment and worker security.
- Upgrading operational effectiveness.

5G likewise can possibly affect industry by dealing with the carbon footprint and narrowing the digital divide. This applies to almost 63 percent of the identified use cases [7].

5G promises to contribute to good health and well-being, advance sustainable industrialization, cultivate innovation, and improve infrastructure. Keeping this in mind, it is said that 5G could be capable of delivering community value over 11 key fields that relate to 11 of the United Nations'17 Sustainable Development Goals. Some other important areas wherein social worth could be created via 5G include empowering manageable cities and networks and advancing work and economic development.

Table 8.5 depicts the use and impact of 5G on three major industries: manufacturing, health care, and mobility.

Apart from the economic and social benefits, 5G offers some environmental benefits. It will lead to cleaner environments by reducing pollution and CO_2 levels and will enhance natural resource management. With the introduction of smart homes, 5G will lead to reduced waste and energy consumption. It will also lead to better and more informed electronic waste.

8.5 5G: HARBINGER OF INDUSTRY 4.0

Industry 4.0 is simply a name given to the present pattern of computerization and transfer of information in the industrialized sector. It incorporates cyber-physical frameworks, the IoT, cloud and cognitive computing, etc. It expects to unite advancements in machines, systems, and offices that originated from the industrial revolution. Industry 4.0 connects factories with value chains that are interconnected globally, and thus it connects providers, clients, colleagues, and others worldwide [8]. Also, frameworks inside the Industry 4.0 paradigm are mindful and provide technological understanding, thus enabling rapid decision-making.

Advantages of Industry 4.0 include the following [9]:

- It allows machines, sensors, gadgets, etc. to be connected and interact with each other.
- Automation and robotics offer indispensable help in situations that seem to be unreasonably perilous for people.

Table 8.5 Impact of 5G in various industrial sectors [7]

S. no.	*Industry sector*	*Use cases*	*Transformation enabled through 5G*
1	Manufacturing	Robotic collaboration with humans Digital performance management Maintenance Augmented reality Virtual reality	Improvement in equipment availability and throughput by using predictive maintenance Operational costs will be reduced with the use of remote maintenance Digital standard operating ways make the process of operations more efficient Manufacturing will become smart for the future factories
2	Health care	Remote patient monitoring Remote robotic surgery Transfer and processing of images Augmented reality Virtual reality health care Disease management Wearable devices Medical deliveries through drone	Access to quality health care is increased with the use of m-Health (mobile health) and telemedicine Health care costs can be reduced by using wearables
3	Mobility	High-density traffic control Automation (C-V2X) Remote monitoring of vehicles' faults Taxis that can move above land	With the introduction of autonomous mobility, individuals will spend less time driving, thus increasing their productivity Environmental impacts will be reduced by green and sustainable mobility

- Creation of a cyber-physical framework empowers frameworks to flawlessly and immediately share information in real time.
- Systems can become self-governing and make on basic choices on their own and continuously.
- Increased technical integration and predictive analytics build efficiency and in this way assist with reducing expenses.
- Companies can offer more customized items that could be profitable and thus extend their business.

Industry cannot rely on the present 3G/4G systems for the features like reliability and latency that are required for the vision of Industry 4.0. 5G is fundamental for the execution of IoT and machine-to-machine communication, which shape the foundation of Industry 4.0, and, henceforth, it is the empowering agent for the vision of Industry 4.0.

5G offers high data rates, less latency, and remarkably high reliability, which are essential to support the Industry 4.0 use cases. These features cannot be offered by the other communication technologies, and, hence, 5G is needed. 5G not only is about consumer mobile broadband but also extends beyond that. 5G will make an enormous impact on industry and mobility. It will permit makers to automate activities and also set up new product lines virtually.

5G will provide flexibility in industrial activities, using its quicker dependable communications between apparatus, sensors, and computing frameworks to establish automated production processes. This will also greatly improve productivity. The reconfiguration of machines in a wireless connectivity environment makes it simpler for manufacturing plants to satisfy processing needs, and cost is also reduced by incorporating wireless tasks. By and large, efficiency and productivity will be increased with automated data and shared information through the production life cycle, empowered with the help of the ultra-low latency and high reliability offered by 5G.

With millions of sensors, machine-controlled robots, etc., all equipped for communication and working in real time through 5G, makers can realize enormous efficiency gains. 5G is going to foster change and lead to development in numerous enterprises, legitimately adding to social and financial advancement [8]. Ultra-low latency along with machine type communication is going to make it simpler for humans and robots to work together. This will also help to keep a check or control on concurrent machines. Furthermore, with massive networks, all the machines will be able to interact with one another, giving constant updates. Machine type communication provides real-time control of machines, robot/human cooperation, and edge cloud analytics. 5G will be critical in implementing the wireless network that will probably control these "smart factories." The key elements in this digitization will be IoT innovation, cloud arrangements, big data, and security. Advancements in 5G can play a key empowering role in coordinating such advances and also offer a platform or a stage on which to interconnect machines, robots, processes, automated vehicles, merchandise, and so on. The job of 5G is to coordinate systems of computing, networking, and storage resources into a single integrated programmable infrastructure. This integration will consider an enhanced, increasingly unique use of distributed resources. The imagined 5G stage should interface the wireless connectivity with the ethernet and incorporate additional segments like edge computing, big data, and analytics.

Maintenance and repairs can be more convenient when 5G empowers virtual/augmented reality communication amongst field staff and industrial facility/product experts. 5G can enable us to have servers on the data centers of the cloud so data upgradation can be done through the cloud without being present in the industrial building, which, in turn, enhances the security of the workforce. To use the total potential that 5G offers to the industrial sector, specialists in every area must team up and bring additional opportunities to the marketplace.

8.6 AUTOMATING 5G

A completely employable and proficient 5G network cannot be finished without consideration of AI. AI and its subclasses like deep learning and machine learning (ML) permit 5G wireless systems to become proactive and predictive, which is fundamental to making the vision of 5G possible. AI is incredible for issues wherein the already present solutions require a great deal of experience and for complex issues having a solution that is not good at utilizing traditional methodologies. It allows us to adjust to altering situations, to get bits of knowledge about complex issues that utilize a lot of data, and also to identify the trends or patterns that a human tends to miss [10]. We have seen AI, mobile, and wireless systems turn into a fundamental social foundation, assemble our everyday lifestyle, and encourage the advanced economy in numerous forms [11]. Nevertheless, in one way or another, AI and 5G wireless communications have been seen as different research fields, in spite of the vast potential they may have if they are intertwined.

Some applications accessible in this convergence of fields have been highlighted within some topics of future or next-generation wireless communication systems and AI. Li et al. [12] featured the ability of AI as an empowering agent for cellular networks to adapt to the 5G normalization prerequisites. Bogale et al. [13] examined ML strategies with regard to edge computing architecture, expecting to disseminate computing power, storage, control, and networking functions. Jiang et al. [14] concentrated on the difficulties of using AI for radio communications in decision making and for adaptive intelligent learning. The upcoming age of wireless and mobile technologies requires the utilization of optimization to maximize or minimize some objective functions [15].

8.6.1 Industry 4.0 use cases enabled by 5G

Automation holds the key to Industry 4.0 and can be classified as follows:

- *Industry automation:* Mechanized activities can create merchandise, which includes gadgets, home appliances, vehicles, and so on.
- *Process control automation:* Procedures can be automatically controlled dependent on ceaseless data acquisition and examination. This applies to oil refineries, power plants, paper factories and so forth.

Industry 4.0 has many use cases dependent on the above-mentioned automation situations. The use cases have their own specifications and necessities as far as information rates, latency, reliability, and so forth are concerned. Some common use cases are as follows [16]:

- *Cell automation:* In this, the devices present in a manufacturing assembly line communicate with the process control system to enhance speed and quality. The prerequisites are latency at least lower than 1 ms and very high consistency.
- *Automated guided vehicles:* These move around within the processing plant, carrying items to various stages as customized. The requirements here are high mobility and reliability.
- *Process automation:* A large number of sensors and actuators are associated with the control units. High reliability is required here.
- *Logistics transportation: The producer keeps track of products all through the supply chain from crude material to delivered merchandise.*
- *Components tracking*: A million static gadgets for each km^2 can be followed at any rate.
- *Augmented reality:* This requires a very high data rate of up to 10 Gbps.
- *Remote robot:* Remote robot control requires extremely high reliability.

8.6.2 5G advances in Industry 4.0

Digitization of industrial procedures has become conceivable because of the diminishing expenses for data storage and computer processing. The biggest development in industrial IoT connections is occurring in the Asia Pacific area. It has been anticipated that about 16.4 billion new industrial IoT connections will be developed worldwide by 2023, and this will significantly increase throughout the following years. As per a McKinsey Global Institute report [17], 75–375 million workers, or 3–14 percent of the worldwide workforce, will be affected by industrial mechanization by 2030 and should change their occupation. Additionally, to improve and upscale the technology, about 50 million new technology workers will be employed by 2030.

Jobs that can be automated will decrease. Jobs involving robotization, specialized jobs, and research and development jobs will increase. Organizations will have the option to make more items with fewer workers. In any case, more individuals will become less fortunate because of job loss and will not have the ability to purchase items. This could prompt a circumstance in which there is overabundance but very little or no demand. This would thoroughly affect social and economic situations everywhere across the world.

8.7 5G NETWORKS ADOPTING AI

According to a report by Ericsson [18], in October–November 2018, it surveyed 165 senior executives from 132 service providers worldwide about the adoption of AI in 5G networks. Their responses indicated that they appreciate AI. They note that ML assists their engineers in making their mobile networks a lot more productive. They also find AI to be important in developing and supporting customer relationships. About 53 percent of those surveyed hope to include AI in their systems before the finish of 2020. A further 19 percent are willing to adopt AI within three to five years.

AI has both advantages and disadvantages for these service providers:

- Is significant for customer service
  Among the service providers surveyed, 55 percent find upgrading the client experience is a key area where AI can be adopted to make the greatest impact within the core network activities. Moreover, 68 percent feature upgrading client service as a business and operational objective for the following three years. Furthermore, 72 percent believe that AI will be significant in empowering the adaptation of new system innovations and offering superior assistance to clients.
- Helps in regaining network investments
  5G has led service providers to make gigantic investments in their systems to empower the new use cases offered by 5G. AI will help operators to recoup these investments more rapidly. The service providers surveyed believe that network planning has the highest potential return (70 percent) from AI adoption. Around 64 percent want to adopt AI as a means to improve network performance.
- Produces network data challenges
  Accommodating AI in systems is a test for the service providers surveyed. The fundamental concern recognized by 71 percent is characterizing and actualizing normalized interfaces. Different challenges featured include data quality (65 percent), overabundance of information from an excessive number of sources (59 percent), issues finding a measure of degradation (59 percent), storage of information in such a large number of systems (56 percent), and absence of a single person or entity to oversee the information (55 percent). At the core of these worries is the way data analytics has been practiced to date using basic and generic tools that will be unsuitable for addressing the huge amount of data in 5G networks.

8.7.1 AI opportunities for service providers: Successful implementation and challenges

5G will provide features such as improved speed, consistency, and reliability. Highlights include, for example, efficient load management that will empower the providers to viably deal with system traffic, guaranteeing its performance does not change with the arrival of more gadgets on the networks. Service providers will have the option to effectively manage network implementation by creating smart, predictive, and a lot more intelligent AI algorithms.

Successful adoption of AI will allow the faster, more responsive available-on-demand networks required for rapid expansion of IoT devices, which need real-time latency. With the expansion of IoT devices in various industries, we need real-time latency. This will be facilitated through AI, which will provide faster and more responsive networks. The service providers are particularly interested in using AI to increase efficiency. Table 8.6 shows the various areas where the service providers are interested in using AI.

The service providers surveyed want to install AI promptly in their systems. This thinking is shared amongst 77 percent worldwide, especially in South East Asia, Oceania, and India (91 percent) and North America (83 percent). These service providers understand that conveyance of new services to clients can be quickened by using AI innovation across systems. Around the world, 76 percent share this view, as do 85 percent in North East Asia. There is a belief among 73 percent internationally that AI will not only make the service providers capable of enhancing their network operations but also allow them to evolve into digital service providers.

To accomplish the combination of AI and 5G, AI innovation should be characterized and implemented utilizing standardized interfaces. AI will without a doubt assume a pivotal role when the network becomes more complex and service providers have to handle several different technologies like 4G, 5G, and IoT, along with the numerous devices that can linked to the network. The reason for this is the ability of AI to enable providers to manage the increase in network deployment and operation costs. In this manner, early adopters of AI-empowered systems will probably profit the most as far as their return on investment is concerned.

To amplify the capability of AI, there should be powerful organization, structuring, and investigation of information to get clear insights into the development and improvement of the network.

The drawbacks of the traditional approach to data analytics can be clearly seen through barriers like low-quality data, privacy concerns, and

Table 8.6 Key areas where service providers will adopt AI [18]

S. no.	*Area*	*Percentage of service providers supporting AI usage*
1	Network capacity and planning	64%
2	Reduced time for network planning	61%
3	Reduced capital expenditures with infrastructure modeling	56%
4	Service quality management	53%
5	Network performance management	53%

complexity. Networks that are enabled by AI use massive data analytics technology, which builds intelligence, event correlation, and adaptiveness in the systems. They can also be helpful in addressing barriers that are preventing action on insights derived from data and also have a role in reducing the associated OPEX.

8.8 5G MULTIPLE-INPUT MULTIPLE-OUTPUT ANTENNA

The frequencies of 5G are suitable for fast data, high capacity, and precise steering control. The demand for mobile communication is increasing with growing technology like cloud computing and mobile computing, so base stations are equipped with multiple-input multiple-output (MIMO) antennas to provide better data transfer and reduced latency. MIMO Antenna focuses energy into a smaller region of space and hence provide better spectral efficiency and throughput.

Recent concerns regarding 5G radiation have arisen. Some scientists believe that it might be fatal to humans, while other scientists oppose this notion. The committee responsible for licensing spectrum—the U.S. Federal Communications Commission—has stated that for a 5G equipment, the signals from the wireless transmitters are typically way below the radiofrequency exposure limits at any location that is accessible to the public.

8.9 EFFECT OF COVID-19 ON 5G

COVID-19 has brutally disrupted the lives of millions of people across the globe. Industries are also suffering a major setback because of it. Several countries are in a lockdown situation, resulting in disruptions to the entire supply chain. The supply of raw materials has become very limited, and as a result, manufacturing suffers. Production facilities have been shut down, reducing the assembly and sales of 5G smartphones. In several countries, the 5G spectrum auction has been delayed by telecommunication authorities, and this will cause a delay in 5G rollout plans and thereby have an effect on the 5G market. It is said that Ericsson and Nokia might feel the greatest impact due to COVID-19, since they have the majority of 5G contracts in the U.S. and European countries.

8.10 CONCLUSION

Worldwide, 5G will create a value of about $3.2 trillion across businesses, with the manufacturing sector representing about one-third of the total. Construction, public services, wholesale and retail, information, and communications will combine to represent another third [2]. To accomplish all of this, trillions will have to be initially spent to establish 5G networks globally, and this gives telecommunication service providers an excellent chance to implement industrial automation. 5G will empower development and change in numerous industries, legitimately adding to socioeconomic development. 5G can open new doors in industries by the creation of employment because of the use of its own system as well as the new applications it makes available. Likewise, we can also expect a loss in jobs in conventional businesses as a result of automation: e.g., robots will be substituted for manual laborers. There will be a need for high-tech professionals. Workforces will have to adjust to the new working condition, since they will be working amongst competent machines. This will require an upgradation of skills for automation staff, though middle- and lower-level employees will suffer.

Presently, with Industry 4.0, nations can completely realize the benefits of 5G by actualizing them more quickly and utilizing them to deliver mass-quality merchandise with fewer workers and with the option to swamp the worldwide market with less expensive products. As production can be robotized, the need to hire work out to low-cost nations might decrease. Both developing nations and poor nations that have inexhaustible amount of unskilled labor will become victims of Industry 4.0 and will also have high jobless rates, prompting social agitation there. The rich nations that can manage the cost of 5G-empowered Industry 4.0 can significantly benefit from new chances.

After investigating a portion of the areas where AI is utilized to enhance 5G advancements, it is firmly believed that the intermingling between these two information technologies will have a colossal effect on the advancement of future networks. The time when wireless systems scientists were hesitant to utilize AI-based calculations because of lack of trust and confidence in AI predictions. These days, with the force and universality of data, various analysts are adjusting their insight and extending their tools to models that are based on AI computations and performance, particularly in the 5G domain, where a couple of milliseconds of latency can have an effect.

8.11 FUTURE SCOPE

Since automation has been done in the core network through AI, an approach to automate and digitalize the base station or the access networks can be developed. The vision of smart base stations making choices without any intervention—creating clusters that are dynamically adaptable dependent on learned information instead of being preset according to rigid principles—will help us improve the productivity, efficiency, and latency of the present systems.

REFERENCES

[1] R. N. Mitra and D. P. Agrawal, "5G mobile technology: A survey," *ICT Express*, vol. 1, no. 3, pp. 132–137, 2015, doi: 10.1016/j.icte.2016.01.003.

[2] IHS Markit, "The 5G economy," November 2019 [Online]. Available: https://www.qualcomm.com/media/documents/files/ihs-5g-economic-impact-study-2019.pdf.

[3] B. Karla and D. K. Chauhan, "A comparative study of mobile wireless communication network: 1G to 5G," *Int. J. Comput. Sci. Inf. Technol. Res.*, vol. 2, pp. 430–433, July 2014.

[4] K. P. Pachauri and O. Singh, "5G technology–Redefining wireless communication in upcoming years," *Int. J. Comput. Sci. Manag. Res.*, vol. 1, no. 1, 2012.

[5] M. G. Kachhavay and A. P. Thakare, "5G Technology—Evolution and revolution," *Int. J. Comput. Sci. Mob. Comput.*, vol. 3, no. 3, pp. 1080–1087, 2014 [Online]. Available: www.ijcsmc.com.

[6] Brahima Sanou, "Setting the scene for 5G: Opportunities and challenges," Telecommunication Development Bureau, ITU, Telecommunication Development Bureau Place des Nations CH-1211 Geneva 20 Switzerland, Published in Geneva, Switzerland, 2018, Persistent link: http://handle.itu.int/11.1002/pub/811d7a5f-en.

[7] World Economic Forum, "The impact of 5G: Creating new value across industries and society," January 2020 [Online]. Available: http://www3.weforum.org/docs/WEF_The_Impact_of_5G_Report.pdf.

[8] W. Haerick et al., "5G and the factories of the future," 5G Infrastructure Public Private Partnership, 2015 [Online]. Available: https://5g-ppp.eu/wp-content/uploads/2014/02/5G-PPP-White-Paper-on-Factories-of-the-Future-Vertical-Sector.pdf.

[9] S. K. Rao and R. Prasad, "Impact of 5G technologies on Industry 4.0," *Wirel. Pers. Commun.*, vol. 100, no. 1, pp. 145–159, 2018, doi: 10.1007/s11277-018-5615-7.

[10] A. Géron, *Hands-On Machine Learning with Scikit-Learn and TensorFlow* (Newton, MA: O'Reilly Media, 2019).

[11] A. Osseiran, J. F. Monserrat, and P. Marsch, eds., *5G Mobile and Wireless Communications Technology* (Cambridge: Cambridge University Press, 2016).

[12] R. Li et al., "Intelligent 5G: When cellular networks meet artificial intelligence," *IEEE Wirel. Commun.*, vol. 24, no. 5, 2017, doi: 10.1109/MWC.2017.1600304WC.

[13] T. E. Bogale, X. Wang, and L. B. Le, "Machine Intelligence techniques for next-generation context-aware wireless networks," *arXiv:1801.04223*, 2018. https://arxiv.org/abs/1801.04223.

[14] C. Jiang, H. Zhang, Y. Ren, Z. Han, K. C. Chen, and L. Hanzo, "Machine learning paradigms for next-generation wireless networks," *IEEE Wirel. Commun.*, vol. 24, no. 2, 2017, doi: 10.1109/MWC.2016.1500356WC.

[15] G. Villarrubia, J. F. De Paz, P. Chamoso, and F. De la Prieta, "Artificial neural networks used in optimization problems," *Neurocomputing*, vol. 272, 2018, doi: 10.1016/j.neucom.2017.04.075.

[16] Ericsson, "Manufacturing reengineered: Robots, 5G and the industrial IoT," *Ericsson Business Review*, 2015. Accessed: October 2, 2018. [Online]. Available: https://www.ericsson.com/assets/local/publications/ ericsson-business-review/issue-4–2015/ebr-issue4-2015-industrialiot.pdf.
[17] J. Manyika et al., "Jobs lost, jobs gained: Workforce transitions in a time of automation," McKinsey Global Institute, December 2017 [Online]. Available: https://www.mckinsey.com/featured-insights/future-of-work/jobs-lost-jobs-gained-what-the-future-of-work-will-mean-for-jobs-skills-and-wages.
[18] Ericsson, "Employing AI techniques to enhance returns on 5G network investments," 2019 [Online]. Available: https://www.ericsson.com/49b63f/ assets/local/networks/offerings/machine-learning-and-ai-aw-screen.pdf.

Chapter 9

Design challenges in planar printed MIMO antennas

Shobhit Saxena, Santanu Dwari, Binod K. Kanaujia, Sachin Kumar, and Mandeep Singh

CONTENTS

9.1 INTRODUCTION

Multiple-input multiple-output (MIMO) wideband/multiband antennas are getting much attention in modern wireless communication systems like ultra-wideband (UWB) and fifth generation (5G). MIMO antennas increase the link reliability of these systems by effectively dealing with multipath fading. Both linearly polarized (LP) and circularly polarized (CP) planar printed MIMO antennas can be designed for contemporary wireless communication applications [1–3]. Sometimes there is a polarization mismatch between transmitter and receiver. In such a scenario, planar printed dual circularly polarized (Dual-CP) MIMO antennas capable of radiating both left-handed circularly polarized (LHCP) and right-handed circularly polarized (RHCP) waves simultaneously will be quite useful. There is a growing

DOI: 10.1201/9781003290230-9

interest in developing wideband and physically small planar printed MIMO antennas fabricated on a single substrate to maintain compactness of the entire system.

The design approaches for planar printed MIMO antennas can be divided into two categories. The first approach makes use of a shared radiator [1], while the second one uses multiple single antenna elements (SAEs) [2].

There are some design issues that must be considered while designing MIMO antennas using multiple SAEs. In this chapter, the design issue of unequal reference voltage levels is highlighted. In some MIMO antennas [4], the ground structures of the SAEs are not connected, leading to unequal reference voltage levels, and, thus, these antennas are of no use in practical scenarios. Besides the above-mentioned design issue, some design challenges related to LP/CP MIMO antennas are discussed in Sections 9.4 and 9.5 of this chapter.

9.2 DESIGN OF MIMO ANTENNAS WITH SHARED RADIATOR

Because significantly less space is available in portable devices, MIMO antennas must be compact. To design compact MIMO antennas, multiple ports can be incorporated in an SAE, leading to the design of a shared radiator MIMO antenna. Figure 9.1 shows an example of a two-port wide-slot antenna having a shared radiator. It is quite challenging to get significant mutual coupling reduction in compact MIMO antennas having a shared radiator [1] compared to those having multiple SAEs [2]. Designing these antennas becomes even more challenging when circular polarization/dual circular polarization [5–7] is required along with mutual coupling reduction.

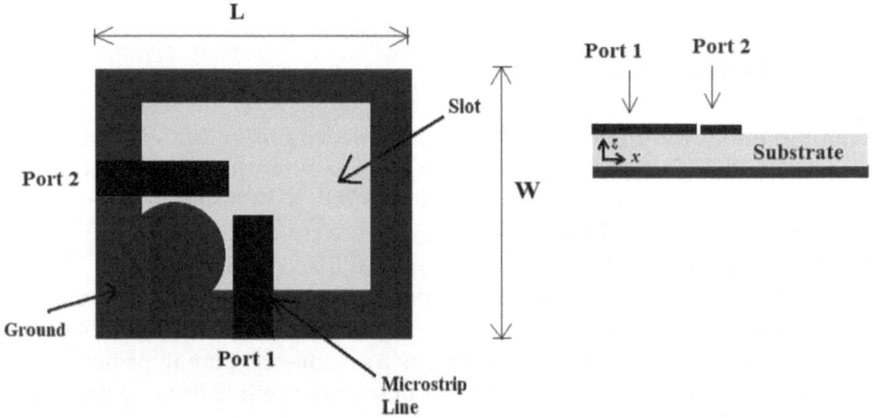

Figure 9.1 An example of a two-port wide-slot antenna having a shared radiator.

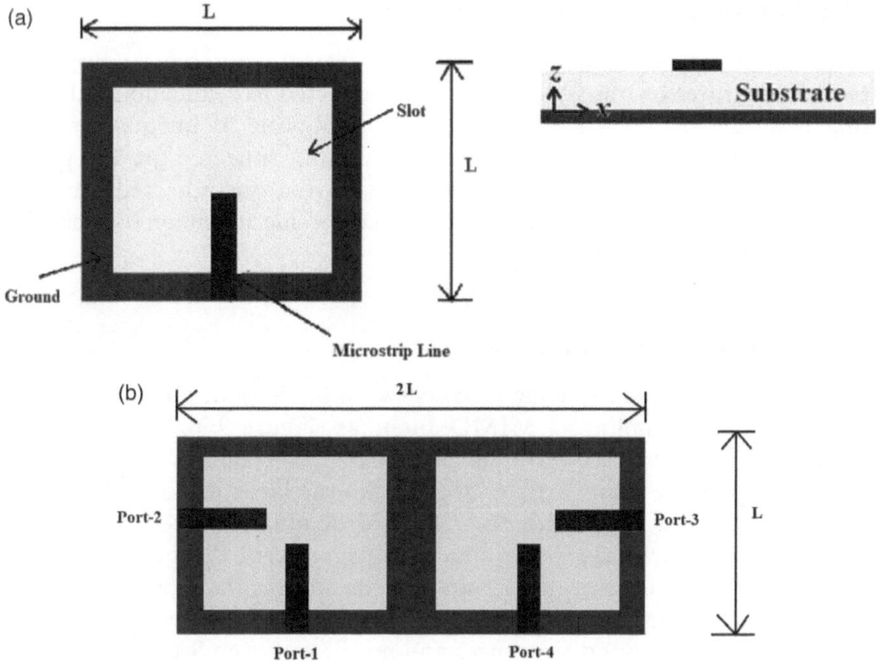

Figure 9.2 (a) An example of a wide-slot SAE. (b) An example of a two-port MIMO antenna designed using SAEs.

9.3 DESIGN OF MIMO ANTENNAS WITH MULTIPLE SAEs

In this approach, multiple SAEs are used for designing multiport MIMO antennas [2]. An example of a two-port MIMO antenna designed using multiple SAEs is shown in Figure 9.2. MIMO antennas having a shared radiator are more compact compared to those using multiple SAEs (Figures 9.1 and 9.2). The main drawback associated with MIMO antennas having a shared radiator is that due to limited available space, only a few ports can be accommodated due to the rise in mutual coupling. On the other hand, there is no such limit on the number of ports in MIMO antennas making use of multiple SAEs.

9.4 DESIGN CHALLENGES IN PLANAR PRINTED LP-MIMO ANTENNAS WITH MULTIPLE SAEs

In this section, some important design challenges related to LP-MIMO antennas with multiple SAEs are discussed.

9.4.1 Unconnected ground planes

In the literature, some articles present LP-MIMO antennas where the ground structures of the SAEs are left unconnected to reduce mutual coupling between SAEs [4, 8, 9] (Table 9.1). This issue of unequal voltage reference levels in MIMO antennas due to the unconnected ground plane is highlighted in [10, 11]. MIMO antennas having unconnected grounds (Figure 9.3) are of no use in practical scenarios due to unequal reference voltage levels in the ground plane.

9.4.2 Interelement spacing

The use of interelement spacing (IES) between SAEs is an effective way of reducing mutual coupling in MIMO antennas (Figure 9.3). It is generally preferred to keep the IES between the SAEs at less than $0.5\lambda_0$ in MIMO antennas. There are many articles available in the literature on LP-MIMO antennas (including those where SAEs are orthogonally placed) where IES is used [2, 12]. The main disadvantage of using extra IES between SAEs is that it increases the overall antenna dimensions. As the world is moving toward miniaturization, some new LP-MIMO antenna designs are required to have connected ground structures and no IES between SAEs.

9.4.3 Additional decoupling structures

To achieve satisfactory mutual coupling reduction, the inclusion of additional decoupling structures between SAEs is a widely used technique in LP-MIMO antennas (Figure 9.3). In the literature, many interesting planar printed LP-MIMO antenna designs making use of additional decoupling structures for reducing mutual coupling between SAEs are proposed [9, 13, 14, 21]. The use of decoupling structures in LP-MIMO antennas has its own advantages and disadvantages. LP-MIMO antennas making use of additional decoupling structures can achieve very high levels of mutual coupling reduction, but there is an increase in the overall complexity and size of the design.

Besides the disadvantage mentioned above regarding the use of additional decoupling structures, there is one more drawback associated with this technique that plays its role at higher frequencies. When MIMO antennas with parasitic elements working as additional decoupling structures are placed in close proximity with other electronic printed circuits operating at comparable frequency ranges, the performance of those circuits will be degraded due to electromagnetic coupling.

LP-MIMO antennas that do not use additional decoupling structures are described in the literature [8, 15, 16]. Although these antennas don't use any decoupling structures, still there is IES present, leading to an increase in their size.

Table 9.1 Planar printed LP-MIMO antennas reported in the literature

Reference	*Year*	*Number of ports*	*Isolation (dB)*	*IES*	*CG*	*Additional decoupling structure used*	*Decoupling technique used*
[2]	2009	2	>16	Yes	Yes	Yes	A tree-like structure is used in the ground plane
[30]	2013	2	>15	Yes	Yes	Yes	Additional stubs are used in the ground plane
[4]	2014	2	>15	Yes	No	Yes	A floating parasitic decoupling structure is used in the ground plane
[8]	2014	2 and 4	>15	Two-port: No Four-port: Yes	Two-port: Yes Four-port: No	No	—
[9]	2015	2	>20	Yes	No	Yes	A rectangular metal strip is placed in between the antenna elements
[13]	2016	2	>15	Yes	Yes	Yes	Two stubs along with CSRR are used in the ground plane
[15]	2016	2	>15	Yes	Yes	No	—
[14]	2018	2	>20	Yes	Yes	Yes	The ground plane has two stubs that look like an F
[31]	2018	2	>20	Yes	Yes	Yes	The ground plane consists of a T-shaped structure
[32]	2018	2	>15	Yes	Yes	Yes	A modified T-shaped stub with a slot is used in the ground plane
[16]	2018	2	>18	Yes	Yes	No	—
[33]	2018	2	>20	Yes	Yes	Yes	Sectioned rectangular slits below the feed line and inverted L-shaped strips on the ground plane are used
[34]	2019	2	>25	Yes	Yes	Yes	An L-shaped parasitic strip and fence type decoupling structure on the ground plane are used
[36]	2013	4	>14	Yes	No	No	—
[37]	2016	4	>22	Yes	Yes	No	—
[38]	2016	4	>20	Yes	Yes	No	—
[39]	2016	4	>22	Yes	No	Yes	Split-ring resonators arranged in the form of a ring are introduced between radiators
[40]	2018	4	>13	Yes	Yes	Yes	The defected ground structure is introduced between antenna elements
[41]	2018	5	>16	Yes	No	No	—
[12]	2019	4	>17	Yes	Yes	Yes	A set of slits is etched in the ground plane
[42]	2020	4	>20	Yes	No	No	—
[17]	2018	4	>11.5	No	Yes	No	—

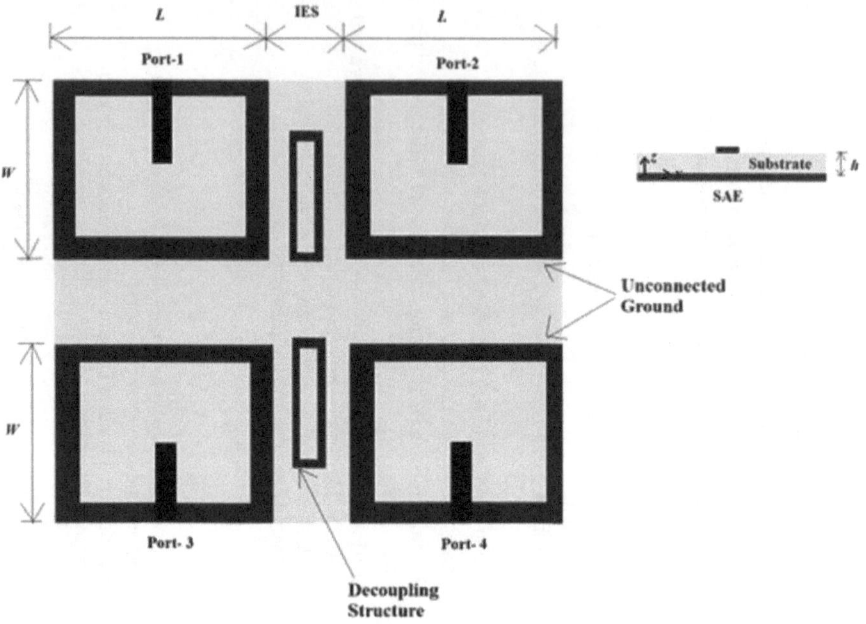

Figure 9.3 An example of a four-port MIMO antenna (having interelement spacing, additional decoupling structure, and unconnected ground) designed using multiple SAEs.

To design simple LP-MIMO antenna structures having no additional decoupling structures and IES, the SAEs need to be designed and arranged in such a manner that they can achieve the minimum required isolation level on their own. In [17], a four-port LP-MIMO antenna is designed using wide-slot SAEs. The wide-slot SAEs are arranged in such a manner that the minimum required isolation level is achieved between the ports of the four-port MIMO antenna without using any additional decoupling structure or IES. It can be clearly seen from Figure 9.4 that the MIMO antenna presented in [17] achieves the minimum required isolation level in the 3.4–3.8 GHz band of interest without using any kind of additional decoupling structure or IES.

9.5 DESIGN CHALLENGES IN PLANAR PRINTED CP/Dual-CP MIMO ANTENNAS WITH MULTIPLE SAEs

In this section, some design challenges related to CP/Dual-CP MIMO antennas are discussed. To facilitate better understanding of the design challenges associated with CP/Dual-CP MIMO antennas, reference is made to a four-port CP-MIMO antenna (antenna-1#1), presented in [18], that includes four identical CP wide-slot-based SAEs.

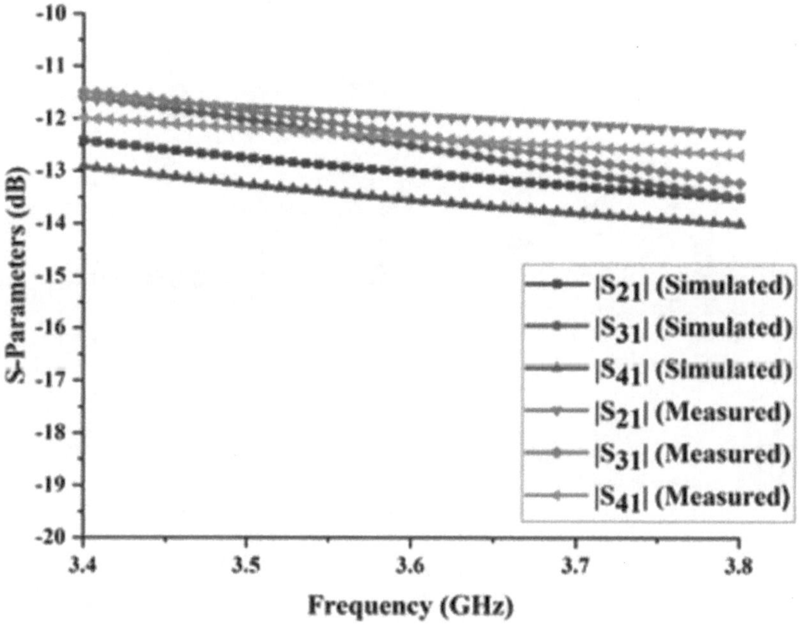

Figure 9.4 Isolation between ports of MIMO antenna (port-1 is excited) presented in [17].

9.5.1 3dB axial ratio bandwidth degradation in closely placed SAEs

In CP/Dual-CP MIMO antennas, additional IES is generally used to restore the CP characteristics of the SAEs [3, 18, 19, 20, 21]. It is usually preferred to keep the IES between SAEs less than $0.5\lambda_0$ in MIMO antennas. If no or significantly less additional IES is used between SAEs, then there is a strong possibility of degradation of the CP characteristics of the SAE even when the ground structures are not connected.

It can be seen from Figure 9.5 that in four-port MIMO antenna-1#1 (without reflector ring), the SAEs are placed near each other, and the ground structures are not connected. Since the SAEs are closely placed, the value of the axial ratio (AR) increases and becomes more than 3dB in the band of interest (5–5.5 GHz), as shown in the axial ratio bandwidth (ARBW) graph in Figure 9.6. From the current distribution of MIMO antenna-1#1(without reflector ring) in Figure 9.7 at 5.2 GHz, it can be seen that this deterioration in AR is due to the strong interference from the adjacent SAEs. Thus, it can be concluded that it is quite challenging to retain the CP properties in MIMO antennas when SAEs are placed very close to each other.

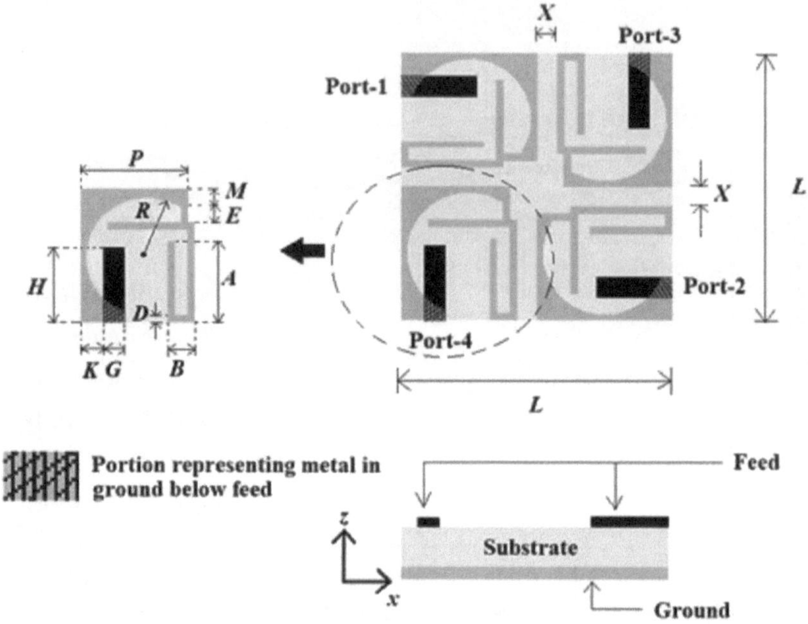

Figure 9.5 Schematic diagram of MIMO antenna-1#1 (without reflector ring).

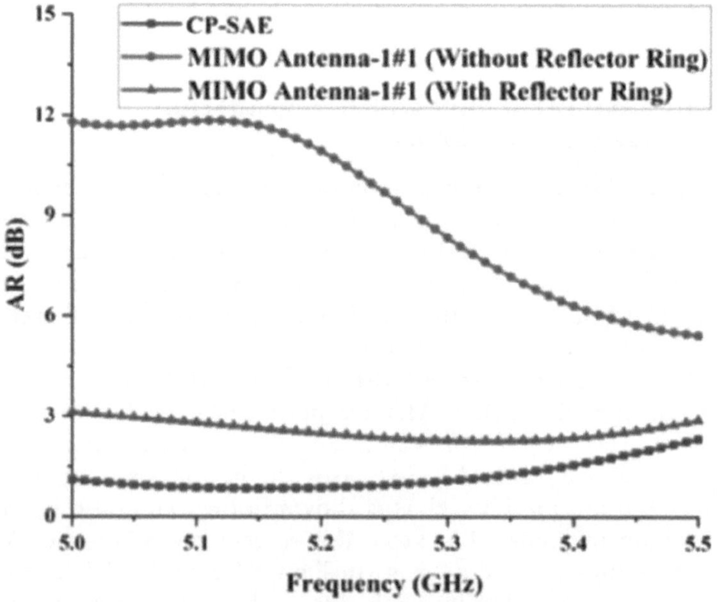

Figure 9.6 ARBW performance of CP-SAE-1#1, MIMO antenna-1#1 (without reflector ring), and MIMO antenna-1#1 (with reflector ring).

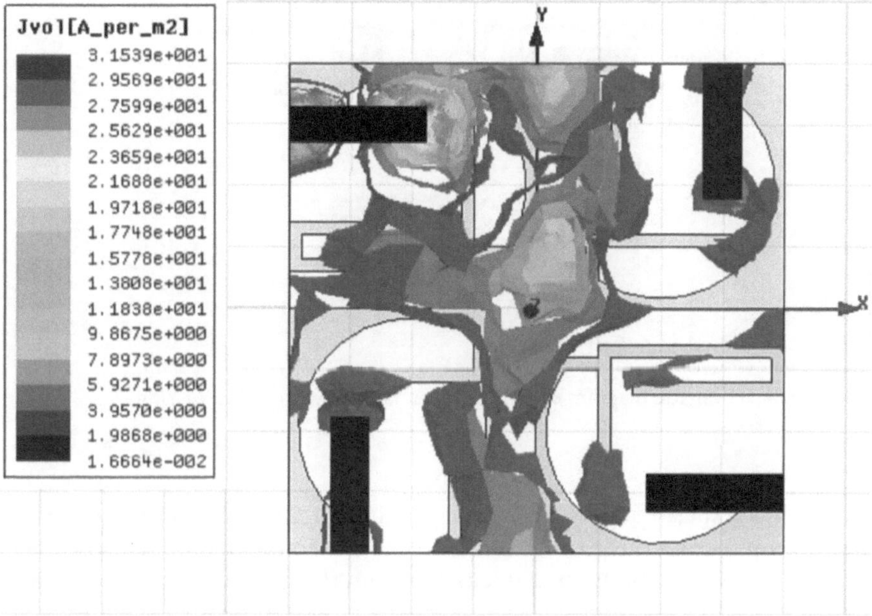

Figure 9.7 Distribution of current at 5.2 GHz in MIMO antenna-1#1 (without reflector ring) (port-1 is excited).

9.5.2 Effect of connecting ground structures on the 3dB ARBW of the MIMO antenna

As discussed in Section 9.4.1, it is important to connect the ground structures of the SAEs in a MIMO antenna to get equal reference voltage levels in the ground plane. This task of connecting the ground structures of the SAEs becomes more challenging in CP-MIMO antennas that use slot- or wide-slot-based SAEs (as the ground is utilized to get circular polarization). The use of additional metallic structures to connect the ground planes of CP-SAEs can degrade the CP performance of the antenna. Hence, the ground structures in CP-MIMO antennas must be connected in such a way that the 3dB ARBW of the SAEs is not disturbed much (in some cases, the 3dB ARBW can be restored in closely placed SAEs by connecting the ground structures, as discussed below).

As an example, MIMO antenna-1#1 (without reflector ring), shown in Section 9.5.1, has CP-SAEs placed very close to each other, thus leading to degradation in the CP performance of the antenna. In MIMO antenna-1#1 (with reflector ring), a circular reflector ring of inner radius R_i = 16.1 mm and outer radius R_o 17.1 mm is introduced in the design for connecting

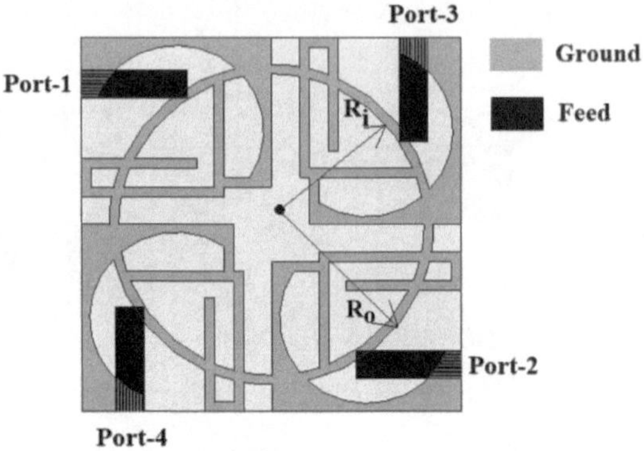

Figure 9.8 A schematic diagram of MIMO antenna-1#1 (with reflector ring).

the ground structures of the CP-SAEs, including CP-SAE-1#1 (Figure 9.8). Besides connecting the ground structures to get equal voltage reference levels, this ring weakens the signal leakage from the adjacent SAEs by reflecting the fields to the radiating antenna element, as can be seen in Figure 9.9. Thus, the 3dB ARBW is restored in MIMO antenna-1#1 (with reflector ring) (Figure 9.6).

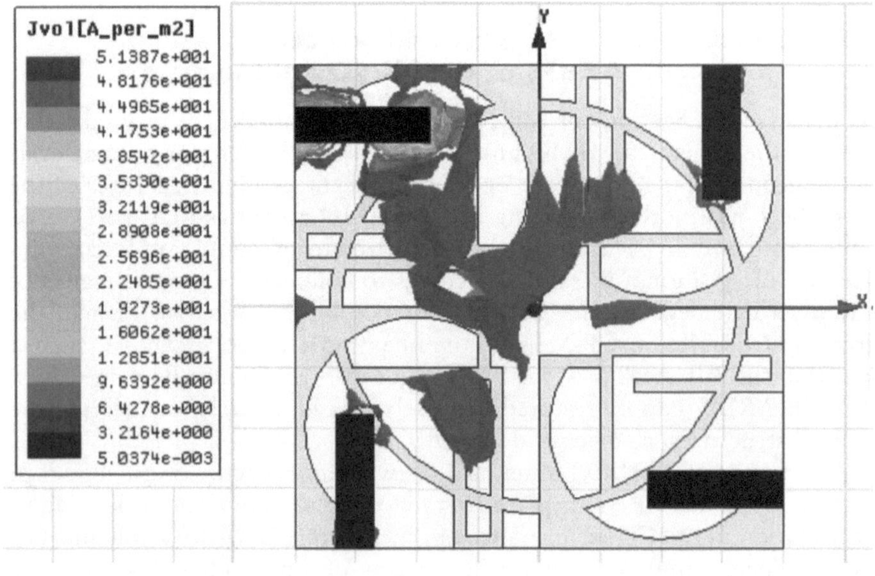

Figure 9.9 Distribution of current at 5.2 GHz in MIMO antenna-1#1 (with reflector ring).

9.6 CONCLUSION AND FUTURE RESEARCH DIRECTIONS

In this chapter, some new challenges related to the design of planar printed MIMO antennas are presented. Some open problems are identified. First, the number of ports can be increased in Dual-CP wide-slot MIMO antennas having a shared radiator without increasing the antenna size. It will be a challenging task to achieve a 3dB ARBW and interport isolation when more than two ports are to be used in the compact wide-slot antennas presented in [6, 23, 26, 27, 35]. The overall dimension of the antennas presented in [22, 24, 25, 28, 29] can be further reduced.

Second, the inter-port isolation in the four-port LP-MIMO antenna presented in [17] can be further improved. Some new wide-slot SAEs can be designed that can achieve better isolation levels in LP-MIMO antennas without using any additional decoupling structure or IES.

Third, there is no article available in the literature on planar printed CP/Dual-CP MIMO antennas that have connected ground structures and no IES. It will be quite challenging to design CP-SAEs in such a manner that the CP characteristics are restored when they are used in CP-MIMO antennas that have connected ground structures and no IES.

Fourth, it is noticed that in most of the articles available in the literature on planar printed CP/Dual-CP MIMO antennas, a very thin metallic strip/structure is used to connect the ground plane. The main reason for using very thin metallic structures is to retain the AR properties of the SAE. It is seen that as the thickness of these metallic structures/strips is increased, the 3-dB ARBW of the SAE deteriorates. Not having enough ground can concentrate the currents at the edges and can causes instability in the response. Thus, there is a need to design CP-MIMO antennas where the SAEs are connected using structures having enough ground.

REFERENCES

[1] Ren J., Hu W., Yin Y., and Fan R. (2014), "Compact printed MIMO antenna for UWB applications," IEEE Antennas and Wireless Propagation Letters, vol. 13, pp. 1517–1520.

[2] Zhang S., Ying Z., Xiong J., and He S. (2009), "Ultrawideband MIMO/diversity antennas with a tree-like structure to enhance wideband isolation," IEEE Antennas and Wireless Propagation Letters, vol. 8, pp. 1279–1280.

[3] Saxena S., Kanaujia B.K., Dwari S., Kumar S., Choi H.C., and Kim K.W. (2020), "Planar four-port dual circularly polarized MIMO antenna for sub-6 GHz band," IEEE Access, vol. 8, pp. 90779–90791. doi:10.1109/ACCESS.2020.2993897.

[4] Khan M.S., Capobianco A.D., Najam A.I., Shoaib I., Autizi E., and Shafique M.F. (2014), "Compact ultra-wideband diversity antenna with a floating parasitic digitated decoupling," IET Microwaves, Antennas and Propagation, vol. 8, no. 10, pp. 747–753.

[5] Saini R.K., and Dwari S. (2016), "A broadband dual circularly polarized square slot antenna," IEEE Transactions on Antennas and Propagation, vol. 64, no. 1, pp. 290–294.
[6] Saxena S., Kanaujia B.K., Dwari S., Kumar S., and Tiwari R. (2017), "A compact dual polarized MIMO antenna with distinct diversity performance for UWB applications," IEEE Antennas and Wireless Propagation Letters, vol. 16, pp. 3096–3099. doi:10.1109/LAWP.2017.2762426.
[7] Kumar A., Ansari A.Q., Kanaujia B.K., and Kishor J. (2019), "Dual circular polarization with reduced mutual coupling among two orthogonally placed CPW-fed microstrip antennas for broadband applications," Wireless Personal Communications, vol. 107, pp. 759–770.
[8] Liu X.L., Wang Z.D., Yin Y.Z., Ren J., and Wu J.J. (2014), "A compact ultra-wideband MIMO antenna using QSCA for high isolation," IEEE Antennas and Wireless Propagation Letters, vol. 13, pp. 1497–1500.
[9] Roshna T.K., Deepak U., Sajitha V.R., Vasudevan K., and Mohanan P. (2015), "A compact UWB MIMO antenna with reflector to enhance isolation," IEEE Transactions on Antennas and Propagation, vol. 63, no. 4, pp. 1873–1877.
[10] Sharawi M.S. (2017), "Current misuses and future prospects for printed multiple-input, multiple-output antenna systems," IEEE Antenna and Propagation Magazine, vol. 59, no. 2, pp. 162–170.
[11] Sharawi M.S. (2013), "Printed multi-band MIMO antenna systems and their performance metrics," IEEE Antennas and Propagation Magazine, vol. 55, no. 5, pp. 218–232.
[12] Villanueva R.G., and Aguilar H.J. (2019), "Compact UWB uniplanar four port MIMO antenna array with rejecting band," IEEE Antennas and Wireless Propagation Letters, vol. 18, pp. 2543–2547.
[13] Khan M.S., Capobianco A.D., Asif S.M., Anagnostou D.E., Shubair R.M., and Braaten B.D. (2016), "A compact CSRR enabled UWB diversity antenna," IEEE Antennas and Wireless Propagation Letters, vol. 16, pp. 808–812.
[14] Iqbal A., Saraereh O.A., Ahmad A.W., and Bashir S. (2018), "Mutual coupling reduction using F-shaped stubs in UWB-MIMO antenna," IEEE Access, vol. 6, pp. 2755–2759.
[15] Lin G.S., Sung C.H., Chen J.L., Chen L.S., and Houng M.P. (2016), "Isolation improvement in UWB MIMO antenna system using carbon black film," IEEE Antennas and Wireless Propagation Letters, vol. 16, pp. 222–225.
[16] Wu Y., Ding K., Zhang B., Li J., Wu D., and Wang K. (2018), "Design of a compact UWB MIMO antenna without decoupling structure," International Journal of Antennas and Propagation, vol. 2018, pp. 1–7.
[17] Saxena S., Kanaujia B.K., Dwari S., Kumar S., and Tiwari R. (2018), "MIMO antenna with built-in circular shaped isolator for sub-6 GHz 5G applications," IET Electronics Letters, vol. 54, no. 8, pp. 478–480. doi: 10.1049/el.2017.4514.
[18] Saxena S., Dwari S., and Kanaujia B.K. (19–22 Dec. 2019, India), "Circularly polarized MIMO antenna with a reflector ring," 2nd IEEE Indian Conference on Antennas and Propagation.
[19] Irene G., and Rajesh A. (2020), "Dual polarized UWB MIMO antenna with elliptical polarization for access point with very high isolation using EBG and MSR," Progress in Electromagnetics Research C, vol. 99, pp. 87–98.

[20] Jamal M.Y., Li M., and Yeung K.L. (2020), "Isolation enhancement of closely packed dual circularly polarized MIMO antenna using hybrid technique," IEEE Access, vol. 8, pp. 11241–11247.
[21] Liu Y.F., Wang P., and Qin H. (2014), "Compact ACS-fed UWB antenna for diversity applications," Electronics Letters, vol. 50, no. 19, pp. 1336–1338.
[22] Mao C.X., and Chu Q.X. (2014), "Compact coradiator UWB-MIMO antenna with dual polarization," IEEE Transactions on Antennas and Propagation, vol. 62, no. 9, pp. 4474–4480.
[23] Zhang J.Y., Zhang F., Tian W.P., and Luo Y.L. (2015), "ACS-fed UWB-MIMO antenna with shared radiator," Electronics Letters, vol. 51, no. 17, pp. 1301–1302.
[24] Khan M.S., Capobianco A.D., Iftikhar A., Asif S., and Braaten B.D. (2016), "A compact dual polarized ultrawideband multiple-input multiple-output antenna," Microwave and Optical Technology Letters, vol. 58, no. 1, pp. 163–166.
[25] Srivastava G., Kanaujia B.K., and Paulus R. (2017), "UWB MIMO antenna with common radiator," International Journal of Microwave and Wireless Technologies, vol. 9, no. 3, pp. 573–580.
[26] Khan M.S., Capobianco A.D., Iftikhar A., Shubair R.M., Anagnostou D.E., and Braaten B.D. (2017), "Ultra-compact dual-polarised UWB MIMO antenna with meandered feeding lines," IET Microwaves, Antennas and Propagation, vol. 11, no. 7, pp. 997–1002.
[27] Patre S.R., and Singh S.P. (2018), "Shared radiator MIMO antenna for broadband applications," IET Microwaves, Antennas and Propagation, vol. 12, no. 7, pp. 1153–1159.
[28] Kaur H., Singh H.S., and Upadhyay R. (19–22 Dec. 2019, India), "Simulation study of quasi self-complementary shared-radiator for UWB-MIMO applications," 2nd IEEE Indian Conference on Antennas and Propagation.
[29] Nie L.Y., Lin X.Q., Xiang S., Wang B., Xiao L., and Ye J.Y. (2020), "High-isolation two-port UWB antenna based on shared structure," IEEE Transactions on Antennas and Propagation. doi:10.1109/TAP.2020.2997461.
[30] Liu L., Cheung S.W., and Yuk T.I. (2013), "Compact MIMO antenna for portable devices in UWB applications," IEEE Transactions on Antennas and Propagation, vol. 61, no. 8, pp. 4257–4268.
[31] Chandel R., Gautam A.K., and Rambabu K. (2018), "Design and packaging of an eye-shaped multiple-input–multiple-output antenna with high isolation for wireless UWB applications," IEEE Transactions on Components, Packaging and Manufacturing Technology, vol. 8, no. 4, pp. 635–642.
[32] Li W., Hei Y., Grubb P.M., Shi X., and Chen R.T. (2018), "Compact inkjet-printed flexible MIMO antenna for UWB applications," IEEE Access, vol. 6, pp. 50290–50298.
[33] Haq M.A., and Koziel S. (2018), "Ground plane alterations for design of high-isolation compact wideband MIMO antenna," IEEE Access, vol. 6, pp. 48978–48983.
[34] Wang L., Du Z., Yang H., Ma R., Zhao Y., Cui X., and Xi X. (2019), "Compact UWB MIMO antenna with high isolation using fence-type decoupling structure," IEEE Antennas and Wireless Propagation Letters, vol. 18, no. 8, pp. 1641–1645.

[35] Li Z., Zhu X., and Zhang J. (15–17 June 2018, China), "Compact dual polarized MIMO antenna for UWB communication applications," 8th International Conference on Electronics Information and Emergency Communication.
[36] Karimian R., Oraizi H., Fakhte S., and Farahani M. (2013), "Novel F-shaped quad-band printed slot antenna for WLAN and WiMAX MIMO systems," IEEE Antennas and Wireless Propagation Letters, vol. 12, pp. 405–408.
[37] Srivastava G., and Mohan A. (2016), "Compact MIMO slot antenna for UWB applications," IEEE Antennas and Wireless Propagation Letters, vol. 15, pp. 1057–1060.
[38] Paramayudha K., Taryana Y., Adipurnama A.B., and Wijanto H. (28–30 July 2016, Indonesia), "Four port diversity patch antennas for MIMO WLAN application," 2016 International Seminar on Intelligent Technology and Its Applications.
[39] Ramachandran A., Mathew S., Rajan V., and Kesavath V. (2016), "A compact tri-band quad element MIMO antenna using SRR ring for high isolation," IEEE Antennas and Wireless Propagation Letters, vol. 16, pp. 1409–1412.
[40] Hussain R., Sharawi M.S., and Shamim A. (2018), "4-element concentric pentagonal slot-line-based ultra-wide tuning frequency reconfigurable MIMO antenna system," IEEE Transactions on Antennas and Propagation, vol. 66, no. 8, pp. 4282–4287.
[41] Nella A., and Gandhi A.S. (2018), "A five-port integrated UWB and narrowband antennas system design for CR applications," IEEE Transactions on Antennas and Propagation, vol. 66, no. 4, pp. 1669–1676.
[42] Islam S.K.N., and Das S. (2020), "Isosceles triangular resonator based compact triple band quad element multi terminal antenna," Radioengineering, vol. 29, no.1, pp. 52–58.

Chapter 10

Four-element dual-band MIMO antenna for Wi-Fi and WLAN applications

Harleen Kaur, Manpreet Kaur, and Hari Shankar Singh

CONTENTS

10.1 INTRODUCTION

User demand for high data rates and speeds has been rapidly growing over the past few years. Systems such as single-input single-output (SISO), single-input multiple-output (SIMO), and multiple-input single-output (MISO) lag in fulfilling the demand for high speed and more channel capacity. However, using a multiantenna system, basically called multiple-input multiple-output (MIMO), can overcome these problems [1]. Thus, MIMO has become the center of attention due to its high-performance characteristics [2]. Many MIMO antenna systems have been proposed so far. However, implementing a MIMO antenna system in a limited area is challenging due to the high risk of generating strong mutual coupling between the radiating elements when closely packed. It degrades the antenna efficiency, provides poor impedance matching, and causes a massive decline in the signal-to-noise ratio (SNR) level [3]. Thus, having a low-profile MIMO antenna

DOI: 10.1201/9781003290230-10

system with good isolation is very important. Some researchers have used techniques to increase the interport isolation. One of the simple ways is to increase the physical separation among elements [4]. However, this increases the overall antenna size; thus, there is a necessity to propose other ways. Many isolation enhancement techniques have been proposed so far for different wireless applications. Further, a neutralization line (NL) has been used to enhance the antenna's interelement isolation [5, 6]. However, connecting the NL with antenna elements can lead to poor impedance matching. Furthermore, researchers have implemented the electromagnetic band gap (EBG) technique to suppress mutual coupling among MIMO array antennas, but it has drawbacks like narrow bandwidth and low gain [7]. The decoupling network at the ground plane has also been used for isolation enhancement among two-element MIMO coradiators for ultra-wideband (UWB) applications [8]. Next, in 2020, [9] proposed a three-port MIMO antenna design in which the isolator also acts as a radiator. However, the large dimensions of the design may limit its applications. Moreover, defected ground structures (DGSs) have been etched to improve the isolation between the radiating elements [10, 11]. A four-port MIMO antenna design has also been proposed for fourth generation (4G) and fifth generation (5G) applications [12, 13]. The proposed design has a large footprint, which can become a major constraint in many applications.

This chapter covers the design and analysis of a four-port dual-band MIMO antenna for Wireless Fidelity (Wi-Fi) and wireless local area network (WLAN) applications. The four antennas (two F-shaped and two elliptical-shaped) are printed on the top side of the substrate, while two slots (partial elliptical-shaped) are loaded on the ground at the bottom layer. The F-shaped antennas are fed with a coaxial probe, whereas the elliptical-shaped antennas are fed by a 50 Ω microstrip line. The simulated results confirm that the proposed design offers high isolation of –16.5 dB over 2.3–2.7 GHz (covering the Wi-Fi band) and –19.5 dB over 5.22–5.64 GHz (covering the WLAN band). Two partial elliptical-shaped slots are etched in the ground to maintain good impedance matching around 5.5 GHz. Further, the parametric analysis is performed to obtain the optimum results and maximum port-to-port isolation. Various diversity parameters are also calculated. The calculated diversity parameters are within the allowable limit for MIMO systems. The 2-D and 3-D radiation patterns are also analyzed. Furthermore, to highlight some of the novelty in the proposed work, Table 10.1 compares the designed MIMO antenna with a few other related antennas. Rest proposed work is divided into different sections; first contains the antenna's configuration and evolution process, followed by its simulated S-parameters and radiation characteristic details. Finally, the chapter is concluded.

Table 10.1 Comparison of the proposed antenna with reference antennas

Reference	*Number of ports*	*Antenna size (mm²)*	*Antenna volume (mm³)*	*Applications*	*Isolation (dB)*	*ECC*	*CCL*
[5]	2	100 × 40	32,000	Diversity	−17	—	—
[6]	2	100 × 40	34,000	UMTS Rx band	−15	—	—
[7]	2	36 × 68	3,916.8	Wireless 5.35 GHz	–43	—	—
[8]	2	29 × 29	673	UWB	−15	<0.005	<0.4
[9]	3	260 × 200	41,600	LTE 2300/LTE 2500/WLAN 2.4/5G C-band	−15	<0.15	—
[10]	4	30 × 35	798	5G mmWave	−17	<0.01	<0.04
[11]	4	12 × 50.8	487.68	5G	–22	—	—
[12]	4	136 × 68	7,398.4	5G	−15	<0.027	—
[13]	4	140 × 75	16,800	4G/5G	–20 at 3.5 GHz and −18 at 4.6 GHz	<0.012	—
Proposed work	4	50 × 110	8,580	Wi-Fi/WLAN	−16.5 at 2.45 GHz and −19.5 at 5.5 GHz	<0.007 and <0.05	<0.13 and <0.39

10.2 ANTENNA CONFIGURATION AND EVOLUTION

This section discusses the design configuration of the proposed four-element MIMO antenna for Wi-Fi and WLAN application. The top and bottom layouts of the proposed structure are shown in Figure 10.1(a). The corresponding simulated reflection coefficients and interport isolation results are depicted in Figure 10.1(b and c). The size of the proposed antenna is equal to $0.4\lambda \times 0.9\lambda \times 0.012\lambda$ mm³, where λ corresponds to the lowest resonant frequency. All four radiators (ANT-1, ANT-2, ANT-3, and ANT-4) have been loaded at the top section of the FR4 substrate, which has a relative permeability (ε_r) equal to 4.4 and a loss tangent (δ) value of 0.02. The antenna has two F-shaped antennas (ANT-1 and ANT-2) at its extremes and two microstrip-fed elliptical-shaped radiating patches (ANT-3 and ANT-4) near its middle section. Each of the identical antenna elements is positioned to behave as the mirror image of another. Further, two coaxial probes and two shorting pins are positioned at the bottom section. The coaxial feed lines excite the F-shaped antennas, whereas the shorting pins connect the F-shaped radiators with the ground plane. In addition, two elliptical-shaped slots are etched in the ground to attain good impedance matching over the WLAN band. The optimized parameters of the proposed

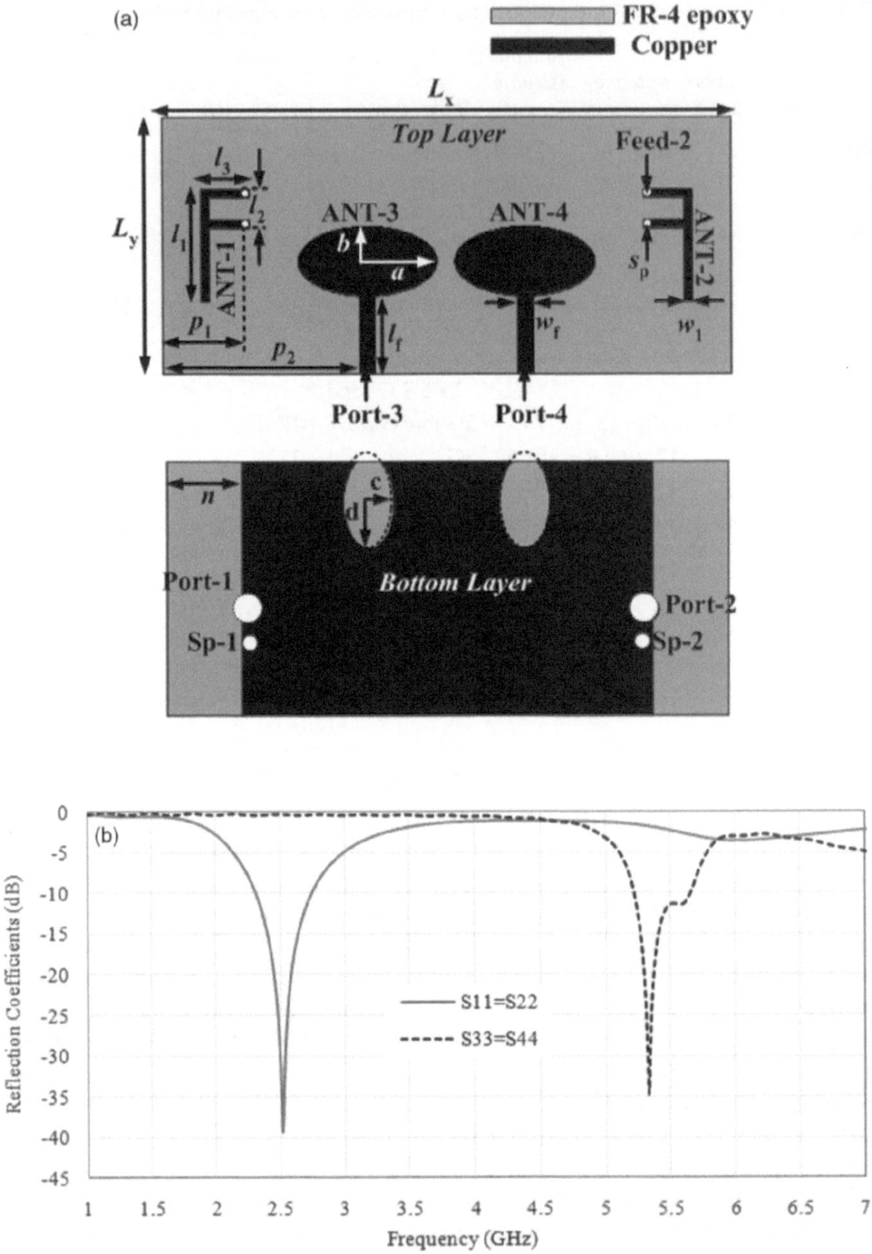

Figure 10.1 (a) Top and bottom layouts of proposed structure; (b) plot of simulated reflection coefficient of proposed radiator; and (c) plot of simulated transmission coefficient of proposed radiator.

(Continued)

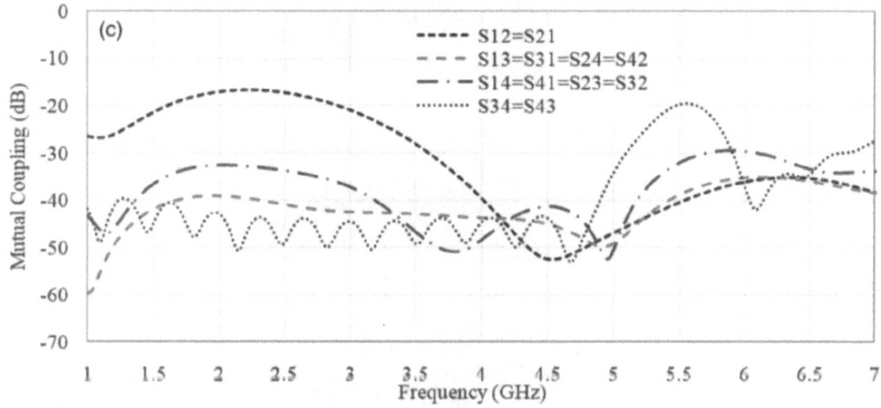

Figure 10.1 (Continued)

radiator are $L_x = 110$ mm, $L_y = 50$ mm, $l_f = 15.3$ mm, $a = 13.5$ mm, $b = 6.75$ mm, $l_1 = 21.8$ mm, $l_2 = 7.4$ mm, $l_3 = 7.92$ mm, $w_1 = 1.48$ mm, $w_f = 1.5$mm, $p_1 = 17.04$ mm, $p_2 = 38.3596$ mm, $c = 3$ mm, $d = 9$ mm, and $n = 15$ mm.

10.3 EVOLUTION OF F-SHAPED ANTENNA

Figure 10.2(a) highlights different designs of the radiators along with the proposed antenna. Its simulated S-parameter results are depicted in Figure 10.2(b and c). The overall dimensions of the ground and its shape are kept fixed to analyze the effect of changing the radiators' shape appropriately. The work starts with the Design-1 configuration, in which two coaxial-fed inverted L-shaped monopole antennas and two microstrip-fed elliptical-shaped radiating elements have been deployed. The simulated reflection coefficients for port-1 and port-2 ($S_{11} = S_{22}$) are under –10dB over frequencies 2.68–2.78GHz (resonate at 2.64 GHz), whereas for port-3 and port-4 ($S_{33} = S_{44}$), they are less than –10 dB over frequencies 5.22–5.63 GHz (resonate at 5.32 GHz). Also, the simulated isolation between antenna elements is more than –16.5 dB over the operating band. The geometry of the inverted L-shaped antennas is almost $\lambda/4$ at 2.6 GHz. Next, two inverted F-shaped antennas are introduced in Design-2 to achieve better impedance matching and shift the resonating frequencies at the lower side (Wi-Fi band of frequencies). This increases the electrical length, but the antennas fail to operate over the Wi-Fi band due to improper impedance. The simulated results show that the reflection coefficients for port-1 and port-2 ($S_{11} = S_{22}$) have poor impedance matching over the desired frequencies, whereas for port-3 and port-4 ($S_{33} = S_{44}$), they are less than –10 dB over frequencies 5.22–5.63 GHz (resonate at 5.32 GHz). Finally, two shorting pins have been used in the proposed radiator to connect each F-shaped antenna element

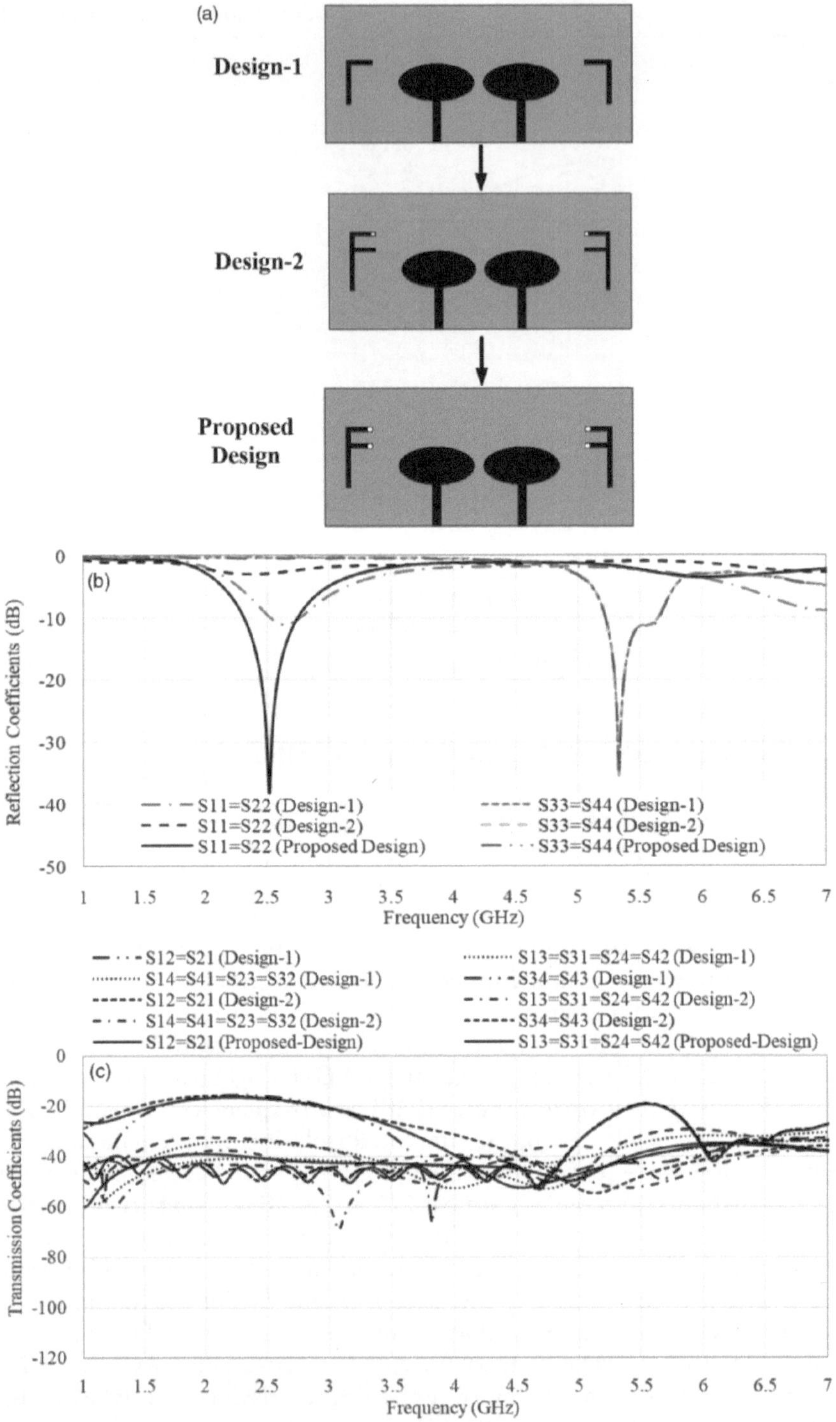

Figure 10.2 (a) Design evolution process of F-shaped antenna; (b) plot curves of simulated reflection coefficient for each design step; and (c) plot curves of simulated transmission coefficient for each design steps.

to the ground plane. By introducing the shorting pins to the conventional inverted F-shaped radiators, good impedance matching over 2.5 GHz has been observed. Thus, after performing a parametric variation on some key design parameters, the proposed four-element MIMO antenna can operate over frequencies 2.3–2.7 GHz (Wi-Fi) and 5.22–5.65 GHz (WLAN).

10.4 EVOLUTION OF ELLIPTICAL-SHAPED ANTENNA

Figure 10.3(a) illustrates the design steps that have been followed to attain the desired dual-band operation in the proposed four-element MIMO antenna. Figure 10.3(b and c) shows the simulated *S*-parameters of each design stage. To highlight the effect more clearly, each design step has a similar ground plane. As seen in Figure 10.3(a), the Step-1 configuration has only F-shaped antenna elements. It is evident from the simulated reflection coefficient results that the Step-1 designed antenna is responsible for generating resonance at the 2.45 GHz. Isolation between the radiating elements is more than –16 dB over the operating band. Next, in order to achieve another wireless operating band, in Step-2 a circular patch antennas of radius r are added. The value of r is chosen such that the electrical length of the structure is approximately $\lambda/4$ at 5.5 GHz. The simulated results show that the Step-2 antenna configuration generates resonance at 2.45 GHz, which mainly occurs because of the presence of the F-shaped antennas. In addition to this, another resonance can be seen at 5.5 GHz, which is due to the circular-shaped antennas. However, as the circular patch antennas offer quite a narrow operating band—i.e., 5.52–5.56 GHz—they are replaced in proposed design by elliptical-shaped patch antennas. Following this replacement, good impedance matching is observed over 5.22–5.64 GHz. It is noted from the simulated *S*-parameters that the proposed radiating structure can now cover both the Wi-Fi and the WLAN bands. Also, the interport isolation at all three steps has shown almost negligible improvement.

10.5 EVOLUTION OF GROUND PLANE

The design methodology of the ground plane is well illustrated in Figure 10.4(a). The effects of different ground plane geometries on the *S*-parameters are shown in Figure 10.4(b and c). Initially, the complete ground plane is analyzed, indicated as GND-1. It has two coaxial probes for feeding and two shorting pins for connecting the antenna elements to the ground plane. However, its operating frequencies do not cover the desired frequency bands. Thus, to improve the input impedance, the conventional ground plane is replaced with a partial ground plane, denoted as GND-2. But the impedance bandwidth lies only within the 2.3–2.7 GHz band (still

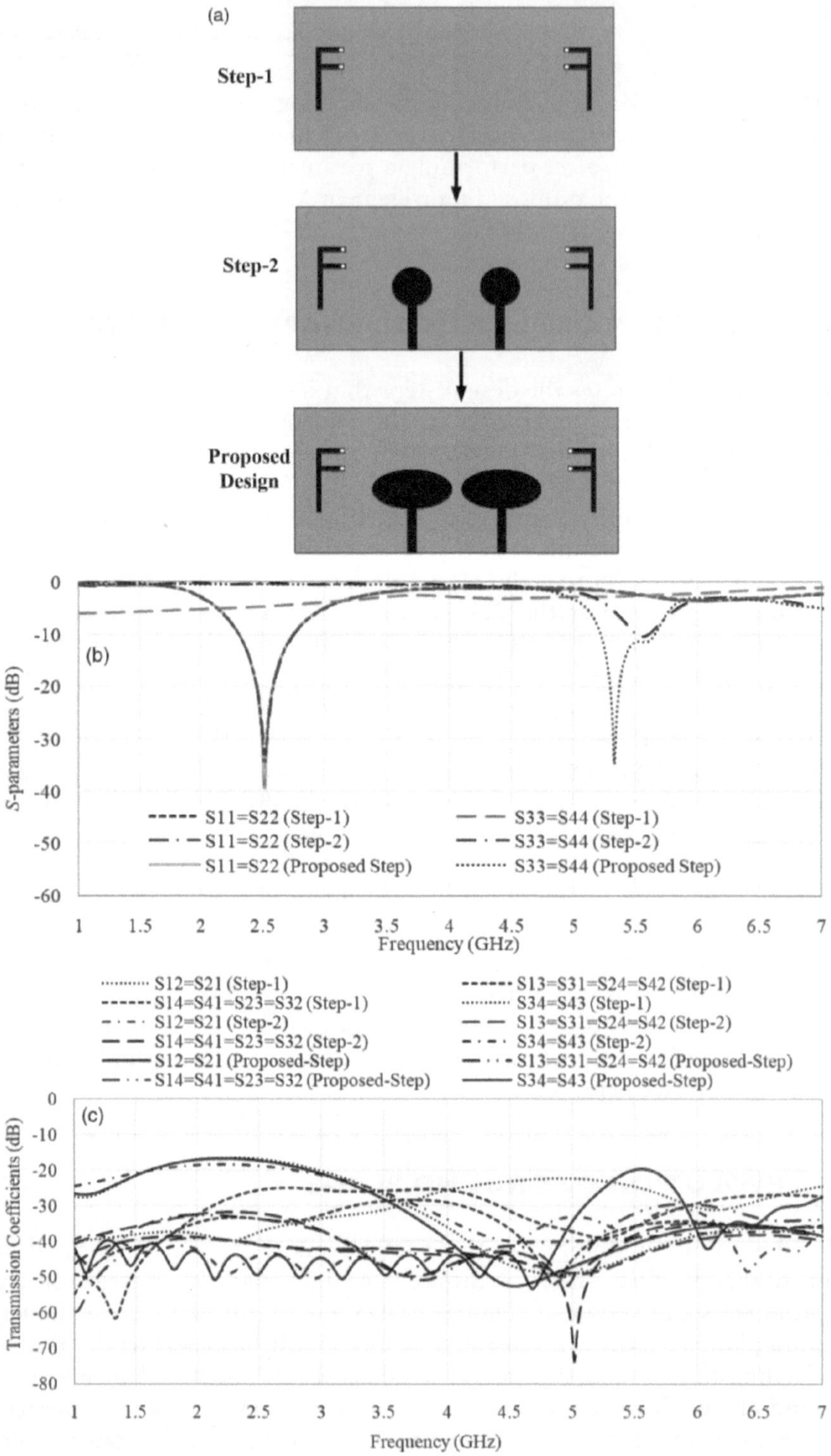

Figure 10.3 (a) Design evolution process of elliptical-shaped antenna; (b) plot curves of simulated reflection coefficient of various design steps; and (c) plot curves of simulated transmission coefficient of various design steps.

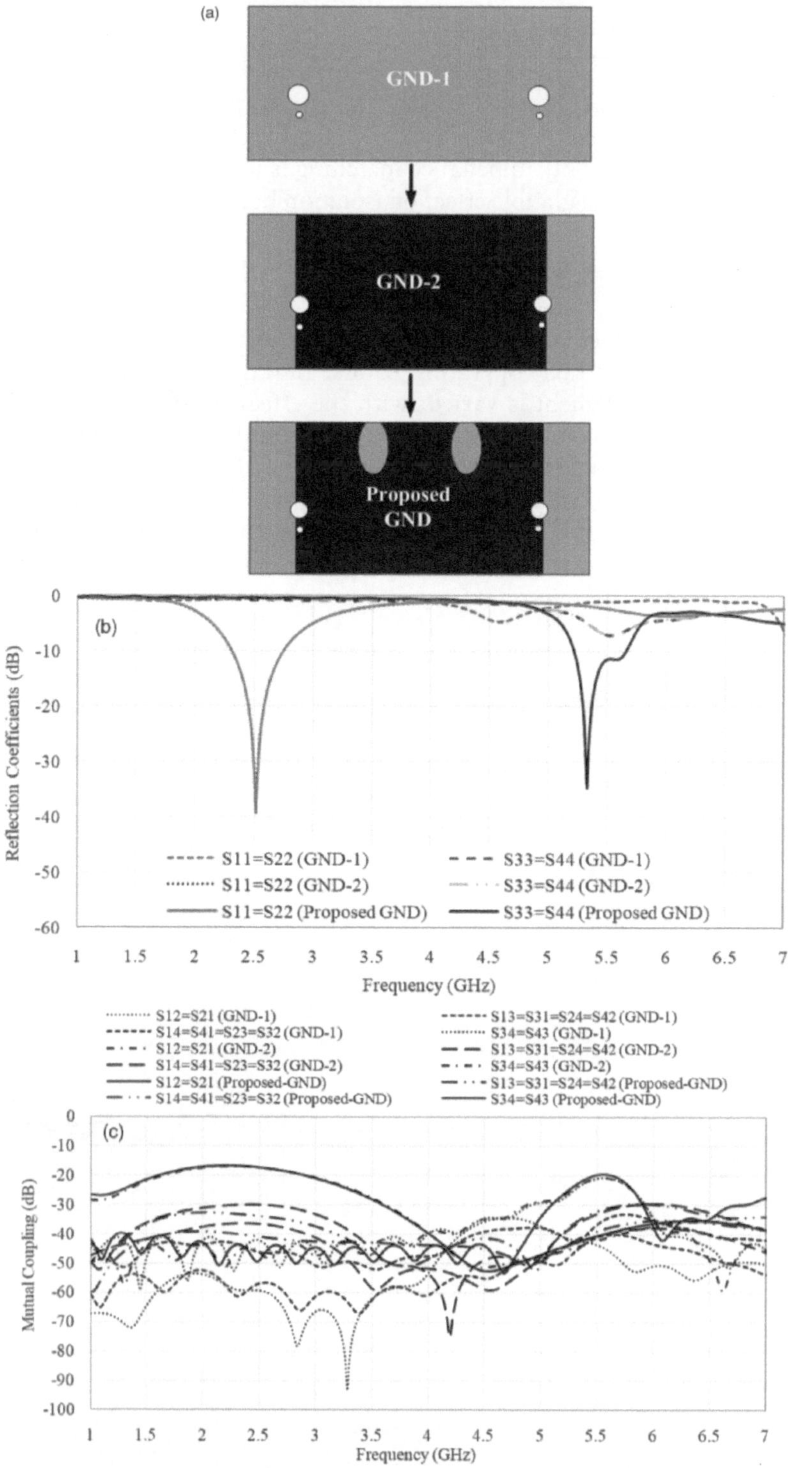

Figure 10.4 (a) Design evolution process of ground plane; (b) plot curves of simulated reflection coefficient of various grounds; and (c) plot curves of simulated transmission coefficient of various grounds.

not covering the WLAN band). Finally, the partial ground plane is further modified by etching two partial elliptical-shaped DGSs. It is noted from the simulated results that only impedance matching is improved with the DGSs. They have an almost negligible effect on isolation between antenna elements.

10.6 PARAMETRIC VARIATION

This section considers some important design parameters that have been optimized to achieve dual operating bands. Initially, the width w_1 of the F-shaped antenna element is varied, with the effect on the reflection coefficients (S_{11}) indicated in Figure 10.5(a). The simulated S_{11} shows that as w_1 increases, the resonance frequency shifts toward the lower side, mainly because the radiating structure's electrical length increases. This parameter affects only the 2.5 GHz operating band because it is part of an F-shaped radiator. After observing the results, the optimized value of w_1 is found to

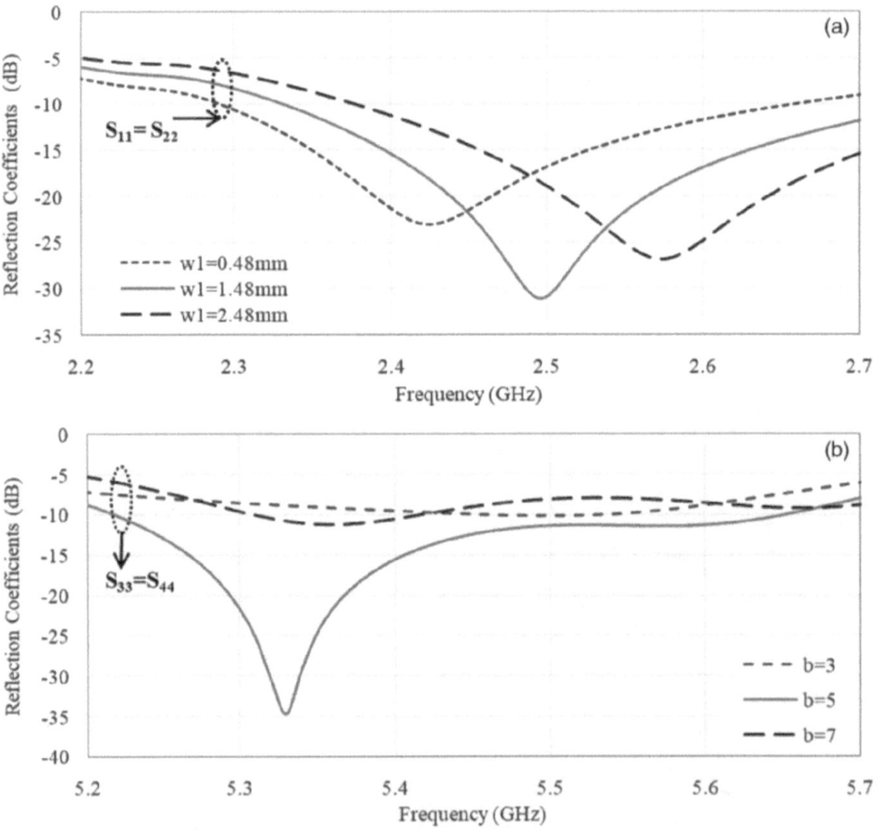

Figure 10.5 (a) Effect of varying 'w_1' on reflection coefficients. (b) Effect of varying slot b on reflection coefficients.

be 1.48 mm. Also, the semiminor axis of partial elliptical-shaped slot b is varied to get the desired operating band. The effect of changing slot b on the reflection coefficients (S_{11}) is plotted in Figure 10.5(b). It is noted that the optimized value is 5 mm, enabling the antenna to attain the desired Wi-Fi and WLAN bands.

10.7 SURFACE CURRENT DISTRIBUTIONS

To validate the proposed MIMO antenna design mechanism, surface currents at two resonating frequencies—i.e., 2.45 and 5.5 GHz—covering all four ports are shown in Figure 10.6. It is observed that at port-1 and port-2 both the F-shaped structures are responsible for the resonance at 2.45 GHz. However, at 5.5 GHz, only the F-shaped structure on the right accounts for

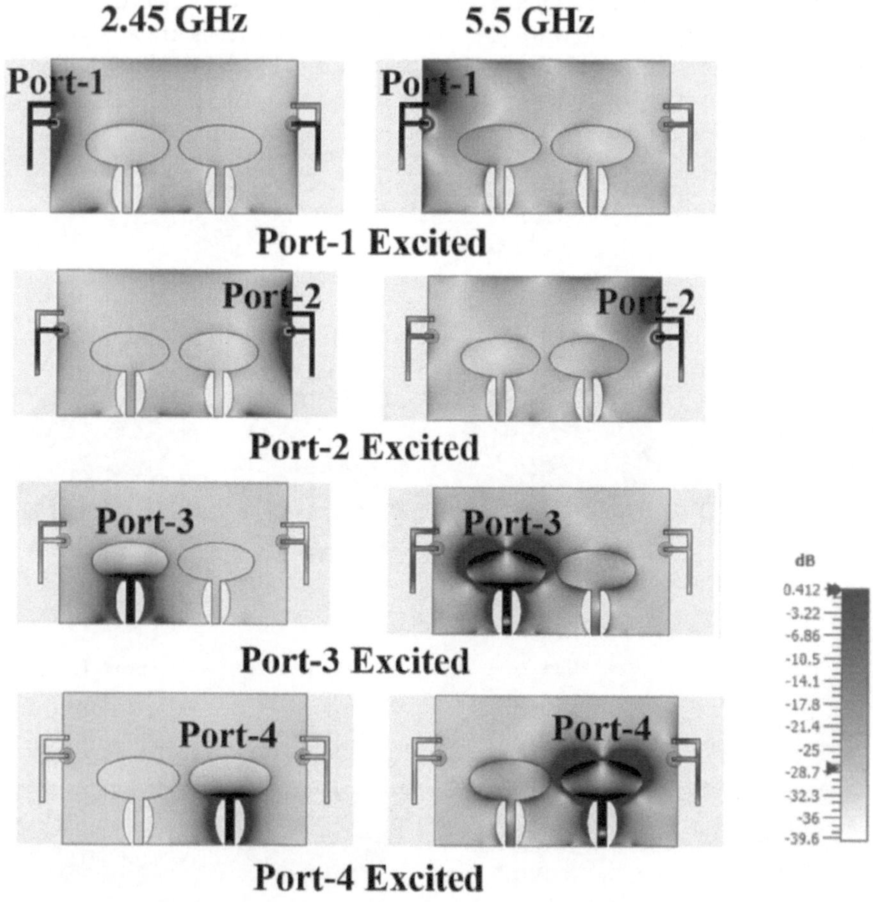

Figure 10.6 Surface current distributions of proposed four-element MIMO antenna at 2.45 and 5.5 GHz.

resonance at port-1, whereas the F-shaped structure on the left creates resonance at port-2. It is interestingly noted that at 2.45 GHz, when only port-3 is excited, the elliptical-shaped slot etched on the left side of the ground plane suppresses most of the surface currents, whereas when port-4 is excited, the elliptical-shaped slot etched on the right suppresses most of the current. Moreover at 5.5 GHz, the elliptical-shaped patch on the left resonates at port-3. Likewise, the elliptical-shaped patch on the right resonates at port-4.

10.8 RESULTS AND DISCUSSION

10.8.1 Radiation performance

Figure 10.7(a and b) depicts 3-D and 2-D far-field radiation patterns of the proposed four-element MIMO antenna at two resonating frequencies: 2.45 and 5.5 GHz. The far-field radiation patterns are obtained using CST MWS electromagnetic software [14]. The radiation patterns are analyzed in two planes—i.e., elevation (E-plane) and azimuth (H-plane). Only a single

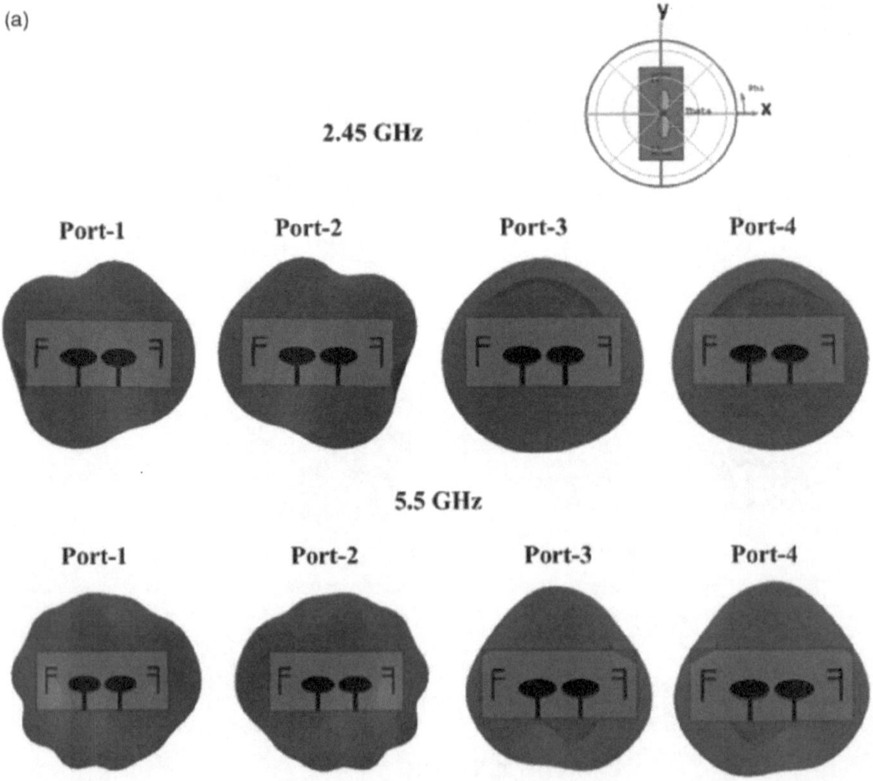

Figure 10.7 (a) 3-D simulated far-field pattern of the proposed antenna. (b) 2-D simulated far-field radiation pattern of the proposed antenna.

(Continued)

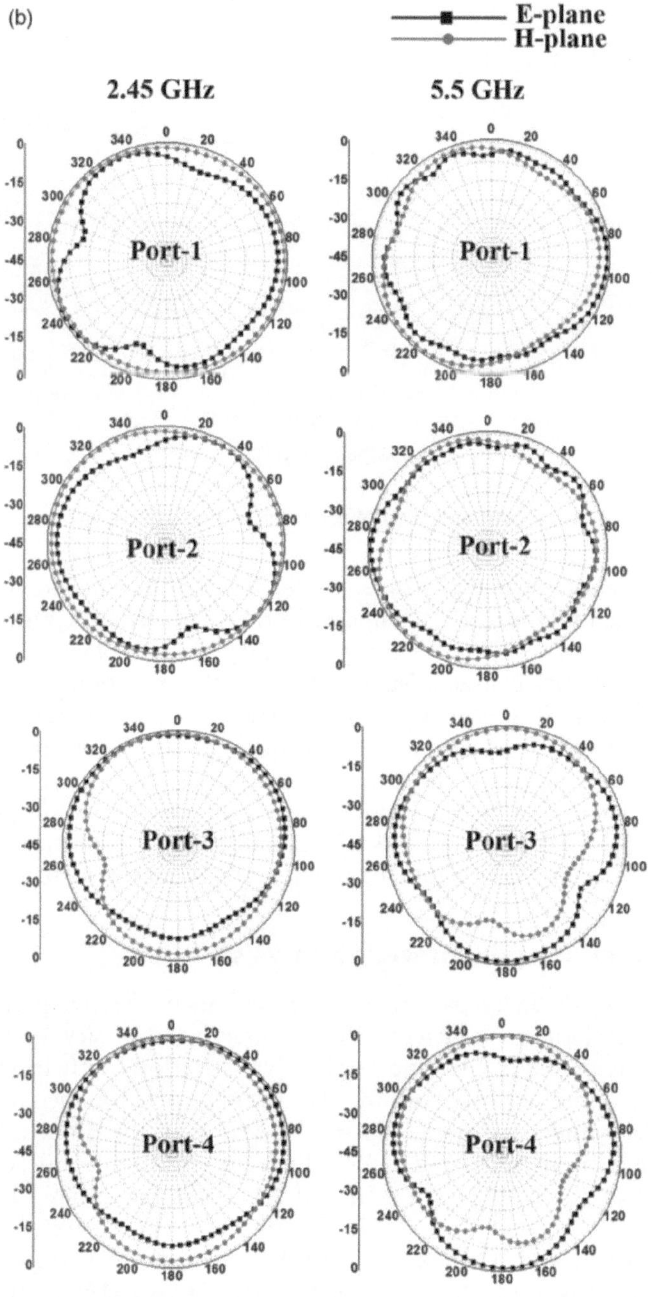

Figure 10.7 (Continued)

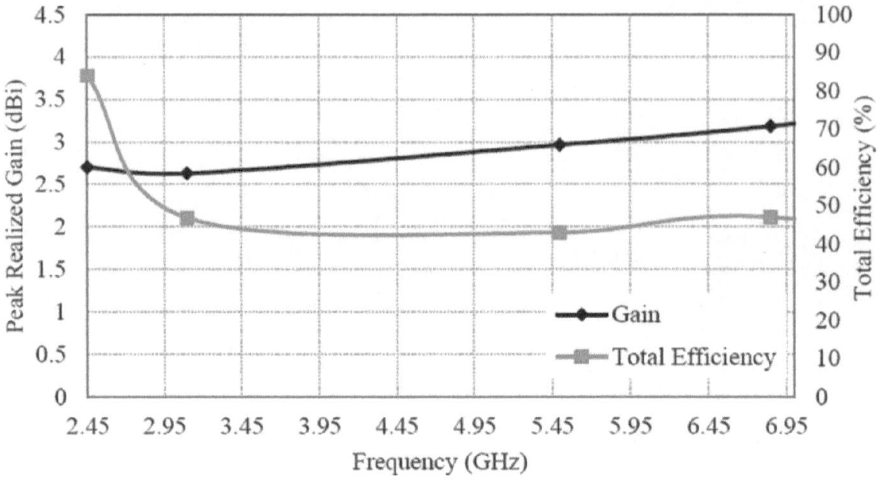

Figure 10.8 Plot of peak gains and total efficiencies of the proposed antenna.

antenna element is excited at a time, while all other elements are terminated with a 50 Ω load. Due to the symmetry in the structures, the radiation patterns of ANT-1 and ANT-2 are complementary in the E-plane, thus exhibiting pattern diversity characteristics. Likewise, the radiation characteristics of ANT-3 and ANT-4 exhibit pattern diversity behavior because they mirror each other's transformation and thus offer good diversity performance. Further, Figure 10.8 shows the simulated peak gains and total efficiencies of the proposed MIMO antenna. It can be seen from the plot that the peak gains at 2.45 and 5.5 GHz occur around 2.7 dBi and 2.9 dBi, respectively, whereas the overall efficiency values are 84 percent and 43 percent, respectively.

10.8.2 Diversity parameters analysis

In this section, diversity parameters such as mean effective gain (MEG), envelope correlation coefficient (ECC), effective diversity gain (EDG), channel capacity loss (CCL), and total active reflection coefficients (TARC) are calculated and used to examine the diversity performance of the proposed MIMO antenna [11–16]. First, the MEGs of the proposed MIMO antenna have been calculated under different environmental conditions, as depicted in Table 10.2. The MEG of ANT-1 is represented as MEG1. Likewise, MEG2, MEG3, and MEG4 are the MEGs corresponding to ANT-2, ANT-3, and ANT-4, respectively. It can be seen from the table that when XPR is 0 dB, the values of MEG1, MEG2, MEG3, and MEG4 are exactly the same. Thus, their ratio is approximately equal to unity. It is also noted that when XPR is 1 dB and 5 dB, the values of MEG1 and MEG2 are nearly the same. Likewise, MEG3 and MEG4 have almost same values with a ratio roughly equivalent to one. This is because a pair of F-shaped antennas will act as a set of individual diversity antenna systems

Table 10.2 MEGs of proposed UWB MIMO antenna

MEG	*Cross-polarization ratio (XPR) (dB)*	*2.45 GHz*
MEG1	Isotropic (XPR = 0)	−3.0103
	Outdoor (XPR = 1)	−3.62342
	Indoor (XPR = 5)	−5.46982
MEG2	Isotropic (XPR = 0)	−3.0103
	Outdoor (XPR = 1)	−3.62341
	Indoor (XPR = 5)	−5.46979
MEGs	*XPR (dB)*	*5.5GHz*
MEG3	Isotropic (XPR = 0)	−3.0103
	Outdoor (XPR = 1)	−4.03583
	Indoor (XPR = 5)	−3.33769
MEG4	Isotropic (XPR = 0)	−3.0103
	Outdoor (XPR = 1)	−4.01475
	Indoor (XPR = 5)	−3.32863

and the partial elliptical-shaped monopole antennas will act as another set, hence satisfying the MIMO equality criteria of diversity systems [15]. Furthermore, ECC has been calculated to measure the correlation among different radiating elements within the MIMO antenna system. Table 10.3 presents the calculated ECC levels of the proposed four-element MIMO

Table 10.3 Diversity parameters of four-port MIMO antenna

Parameter	*2.45 GHz*
ECC for port-1 and port-2 ($\rho_{12} = \rho_{21}$)	0.01
ECC for port-1 and port-2 ($\rho_{13} = \rho_{31} = \rho_{24} = \rho_{42}$)	0.003
ECC for port-1 and port-2 ($\rho_{14} = \rho_{41} = \rho_{23} = \rho_{32}$)	0.07
CCL for port-1 and port-2	0.13
TARC for port-1 and port-2 at phase diff. = 0	0.016
TARC for port-1 and port-2 at phase diff. = 60	0.07
TARC for port-1 and port-2 at phase diff. = 120	0.02
TARC for port-1 and port-2 at phase diff. = 180	0.03
EDG for port-1 and port-2	8.2
Parameter	*5.5 GHz*
ECC for port-3 and port-4 ($\rho_{34} = \rho_{43}$)	0.001
ECC for port-3 and port-4 ($\rho_{13} = \rho_{31} = \rho_{24} = \rho_{42}$)	0.002
ECC for port-3 and port-4 ($\rho_{14} = \rho_{41} = \rho_{23} = \rho_{32}$)	0.005
CCL for port-3 and port-4	0.39
TARC for port-3 and port-4 at phase diff. = 0	0.18
TARC for port-3 and port-4 at phase diff. = 60	0.04
TARC for port-3 and port-4 at phase diff. = 120	0.19
TARC for port-3 and port-4 at phase diff. = 180	0.25
EDG for port-3 and port-4	6.8

antenna. The simulated ECC values are not more than 0.5 at the operating regions, which is a requirement of any diversity antenna system [16]. Moreover, the EDGs of the proposed antenna at 2.45 and 5.5 GHz are analyzed, as presented in Table 10.3. Mathematically, the apparent diversity gain (DG) [17] is computed from Equation (10.1):

$$DG_{apparent} = 10\sqrt{1 - ECC} \tag{10.1}$$

Then to obtain EDG, the apparent DG is multiplied by total antenna efficiency. After that, TARC, another crucial diversity parameter used to evaluate MIMO antenna systems' performance, is calculated [18, 19]. The generalized equation used to compute TARC for an N-port antenna is given in Equation (10.2):

$$T_a^t = \sqrt{\sum_{i=1}^{N} |b_i|^2 \Big/ \sum_{i=1}^{N} |a_i|^2} \tag{10.2}$$

$$\begin{bmatrix} b_1 \\ b_2 \\ b_3 \\ b_4 \end{bmatrix} = \begin{bmatrix} S_{11} & \cdots & S_{14} \\ \vdots & \ddots & \vdots \\ S_{41} & \cdots & S_{44} \end{bmatrix} \begin{bmatrix} a_1 \\ a_2 \\ a_3 \\ a_4 \end{bmatrix} \tag{10.3}$$

where a_i and b_i denote the incident and reflected signals, respectively. In the case of the four-port MIMO antenna, the relationship between the scattering and the incident and reflected signals is depicted in Equation (10.3), where [*S*] represents the scattering matrix, [a] is the excitation array vector, and [b] denotes the scattered array vector. The value of TARC should always lie between 0 and 1. When the value of TARC is 0, it indicates that the maximum power is radiated. When the value of TARC is 1, it indicates that all the power either gets reflected or is transferred into other ports. The variation of TARC with different phase angles between the antenna elements of the proposed MIMO antenna is presented in Table 10.3. Furthermore, to check the performance of the MIMO system, CCL is analyzed. CCL values should ideally be as low as 0 but not more than 0.4 bits/Hz/sec in the case of a suitable wireless communication environment [20]. The proposed antenna offers CCL within the desired limit for all antenna elements, as given in Table 10.3; hence, the proposed antenna has the potential to provide stable diversity performance.

10.9 CONCLUSION

In this chapter, a four-port MIMO antenna for Wi-Fi and WLAN applications is presented. A pair of F-shaped MIMO antennas, along with another pair of elliptical-shaped patch antennas, has been used to attain the desired frequency bands. The simulated results show that the mutual coupling between elements is more than –16.5 dB between port-1 and port-2, whereas it is –19.5 for port-3 and port-4 without using any isolation enhancement technique. Further, the 2-D and 3-D simulated radiation patterns plots confirm that the proposed antenna exhibits good pattern diversity performance over the operating bands. Also, the simulated ECC is not more than its defined limit (i.e., < 0.5), which fulfills the criterion of diversity antenna systems. Furthermore, TARC and CCL are calculated and found to be within acceptable limits over the desired frequency range. Based on this thorough analysis, it is found that the proposed antenna has sufficiently high interport isolation and is well suited for Wi-Fi and WLAN MIMO applications.

ACKNOWLEDGMENT

The authors would like to take an opportunity to thank the Department of Electronics and Communication Engineering, Thapar Institute of Engineering and Technology, Patiala, Punjab, India, for providing the laboratory and the seed money grant to carry out this research. The authors also express gratitude to M/s HAMA IoT Solutions Private Ltd. Sonbhadra, Uttar Pradesh (incubated at TIDES, IIT Roorkee under DST NIDHI scheme PRAYAS) for the support to carry out this work.

REFERENCES

[1] Foschini, G. J., and Gans, M. J. (1998). On limits of wireless communications in a fading environment when using multiple antennas. *Wireless Person. Commun.*, 6, 311–335.

[2] Iqbal, A., Altaf, A., Abdullah, M., Alibakhshikenari, M., Limiti, E., and Kim, S. (2020). Modified U-shaped resonator as decoupling structure in MIMO antenna. *Electronics*, 9(8), 1321 (1–13). DOI: 10.3390/electronics9081321.

[3] Friel, E. M., and Pasala, K. M. (2000). Effects of mutual coupling on the performance of STAP antenna arrays. *IEEE Trans. Aerosp. Electron. Syst.*, 36(2), 518–527. DOI: 10.1109/7.845236.

[4] Lee, H. M., and Lee, H. (2012). Isolation improvement technique for two closely spaced loop antennas using MTM absorber cells. *International Journal of Antennas and Propagation*, 1–9. DOI: 10.1155/2012/736065.

[5] Diallo, A., Luxey, C., Thuc, P. L., Staraj, R., and Kossiavas, G. (2008). Enhanced two-antenna structures for universal mobile telecommunications system diversity terminals. *IET Microwaves Antennas Propag.*, 2(1), 93–101. DOI: 10.1049/iet-map:20060220.

[6] Chebihi, A., Luxey, C., Diallo, A., Thuc, P. L., and Staraj, R. (2008). A novel isolation technique for closely spaced PIFAs for UMTS mobile phones. *IEEE Antennas Wireless Propag. Lett.*, 7, 665–668. DOI: 10.1109/LAWP.2008.2009887.

[7] Moghadasi, M. N., Ahmadian, R., Mansouri, Z., Zarrabi, F. B., and Rahimi, M. (2014). Compact EBG structures for reduction of mutual coupling in patch antenna MIMO arrays. *Prog. Electromagn. Res. C*, 53, 145–154. DOI: 10.2528/PIERC14081603.

[8] Kaur, H., Singh, H. S., and Upadhyay, R. (2021). A compact dual-polarized co-radiator MIMO antenna for UWB applications. *Int. J. Microwave Wireless Technolog.*, 1–14. DOI: 10.1017/S1759078721000349.

[9] Chen, W. S., and Lin, R. D. (2021). Three-port MIMO antennas for laptop computers using an isolation element as a radiator. *Int. J. RF Microwave Comput.-Aided Eng.*, 31(2), 1–11. DOI: 10.1002/mmce.22326.

[10] Khalid, M., Naqvi, S. I., Hussain, N., Rahman, M., Fawad, Mirjavadi, S. S., Khan, M. J., and Amin, Y. (2020). 4-port MIMO antenna with defected ground structure for 5G millimeter wave applications. *Electronics*, 9(1), 71(1–13). DOI: 10.3390/electronics9010071.

[11] Jilani, S. F., and Alomainy, A. (2018). Millimetre-wave T-shaped MIMO antenna with defected ground structures for 5G cellular networks. *IET Microwaves Antennas Propag.*, 12(5), 672–677. DOI: 10.1049/iet-map.2017.0467.

[12] Abdullah, M., Ban, Y. L., Kang, K., Sarkodie, O. K. K. F., and Li, M. Y. (2017, March 26–30). Compact 4-port MIMO antenna system for 5G mobile terminal. *IEEE International Applied Computational Electromagnetics Society Symposium*, Firenze, Italy.

[13] Biswas, A., and Gupta, V. R. (2020). Novel compact planar four-element MIMO antenna for 4G/5G applications. *Nanoelectron. Circuits Commun. Syst.*, 692, 109–117. DOI: 10.1007/978-981-15-7486-3_12.

[14] CST Microwave Studio (2016). A numerical simulation software for electromagnetic computing. *Computer Simulation Technology GmbH, Darmstadt, Germany*, https://ww.cst.com/.

[15] Taga, T. (1990). Analysis for mean effective gain of mobile antennas in land mobile radio environments. *IEEE Trans. Veh. Technol.*, 39, 117–131. DOI: 10.1109/25.54228.

[16] Singh, H. S., Pandey, G. K., Bharti, P. K., and Meshram, M. K. (2015). Design and performance investigation of a low profile MIMO/diversity antenna for WLAN/WiMAX/HIPERLAN applications with high isolation. *Int. J. RF and Microwave Comput.-Aided Eng.*, 25(6), 510–521. DOI: 10.1002/mmce.20886.

[17] Kaur, H., Singh, H. S., and Upadhyay, R. (2021). Design and experimental verification of compact dual-element quasi-self-complementary ultra-wideband multiple-input multiple-output antenna for wireless applications. *Microwave Optical Technol. Lett.*, 63(6), 1774–1780. DOI: 10.1002/mop.32819.

[18] Andrade, E. F., Aguilar, H. J., and Mendez, J. A. T. (2019). The correct application of total active reflection coefficient to evaluate MIMO antenna systems and its generalization to N ports. *Int. J. RF and Microwave Comput.-Aided Eng.*, 30(5), e22113 (1–10). DOI: 10.1002/mmce.22113.
[19] Chae, S. H., Oh, S. K., and Park, S. O. (2007). Analysis of mutual coupling, correlations, and TARC in WiBro MIMO array antenna. *IEEE Antenna Propag. Lett.*, 6(11), 122–125. DOI: 10.1109/LAWP.2007.893109.
[20] See, C. H., Alhameed, R. A. A., Abidin, Z. Z., McEwan, N. J., and Excell, P. S. (2012). Wideband printed MIMO/diversity monopole antenna for WiFi/WiMAX applications. *IEEE Trans. Antennas Propag.*, 60(4), 2028–2035. DOI: 10.1109/tap.2012.2186247.

Chapter 11

MIMO antenna applications

RFID, WLAN, wearable antennas, and IoT

Durga Prasad Mishra, Biswajit Dwivedy, Tanmaya Kumar Das, and Santanu Kumar Behera

CONTENTS

DOI: 10.1201/9781003290230-11

11.1 INTRODUCTION

The efficiency of a wireless network becomes a concern with the increase in the number of end users and antennas on communication equipment, which thereby also increases the huge multiuser MIMO (MU-MIMO) network output. The number of wireless gadgets intended to be worn on the human body has risen enormously in recent years because these wearable devices affected our lives in many ways. By communicating with the human body for health screening reasons, MIMO can be effective as wearable technology that detects vital signs such as heart rate and blood pressure. This is achieved by deploying multiple antennas as transceivers, which significantly eliminates fading. For use in portable devices, MIMO antenna systems need high decoupling between antenna ports and a small scale. The transmitting and receiving ends have multiple antennas. The single-input multiple-output (SIMO) and multiple-input single-output (MISO) technologies can be integrated into a single device to increase the efficiency of a MIMO system. The use of MIMO can increase channel capacity by spatial multiplexing (SM) and diversity by largely eliminating fading with a low power requirement.

A chipless radio-frequency identification (RFID) system [1–7] with MIMO antenna technology is the best candidate to satisfy the demand for low-cost devices. It uses the SM capacity and high diversity of low-cost retail and health care applications. This is one of the suitable techniques used in advanced mobile robot–based MIMO systems, synthetic aperture radar (SAR) based RFID localization, electromagnetic (EM) imaging in RFID systems, and internet of things (IoT) applications [2]. This chapter describes various applications of MIMO systems in the RFID field as advanced non-line-of-sight technology without human intervention [1–2]. It also explains the use of Wireless Fidelity (Wi-Fi) to minimize issues raised by a multipath channel under the IEEE 802.16e standard that implements MIMO-based orthogonal frequency-division multiple access (OFDMA). The new advances in antenna technologies include the super-wideband MIMO antenna for IoT applications. In 802.11ac, the MU-MIMO relays separate data streams concurrently to various users in a single frequency band by using more antennas by the access point (AP) [2–5]. This chapter also describes several designs and implementation problems from the viewpoint of the reader [8]. A smart application of the MIMO framework is discussed in this chapter. The output of many wearable MIMO systems consisting of single-layer and multiple-layer wearable electronic textile antennas, which are available at various locations, are taken into account for throughput evaluation. The semi-direct radiation characteristics of the fabric patch antenna are preferable for wearable applications to eliminate the need for excessive sensitivity to human radiation in a wireless body area network (WBAN). MIMO can be suitably implemented for firefighters working in big infrastructure buildings who communicate by using multiple wearable textile patch antennas in the fabric of their clothing.

This chapter addresses various applications of a MIMO system in the fields of RFID, the wireless local area network (WLAN), WBAN, and

IoT, noting the key points from an antenna and propagation perspective to enable efficient MIMO communications.

11.2 RFID PRINCIPLE OF OPERATION

RFID technology has advanced from obscurity to mainstream applications in recent years, allowing for quicker handling of manufactured products and materials. RFID allows for detection from a distance, and unlike existing bar-code technology, it does so without having a line of sight. There are several different forms of RFID, but at the most basic level, RFID devices can be divided into two categories: active and passive. Active tags need to be powered, so they're either connected to a power source or dependent on energy stored in a built-in battery. The lifespan of a tag is limited in the latter case by the stored energy, which is balanced against the number of reading operations the system must perform. Active tags are impractical for the retail trade due to their expense, size, and life span. In passive tags, there is no on-board power source or active transmitter. Here, the RFID reader's EM signal inductively drives the tag, allowing it to retransmit its data. The tags are also small enough to fit into a practical adhesive label and have an indefinite operating life. Figure 11.1 shows the architecture of a basic RFID system, which consists of a reader (or interrogator) subsystem, tag (or transponder) subsystem, and middleware to control the reader and collect data [2]. The power is supplied to the passive tag by the reader unit by transmitting EM waves. The tag is a passive device without any integrated circuitry, so it reflects the received wave toward the reader, which is known as backscattering. The backscattered wave is received by the reader antenna and sent for further processing by different hardware and middleware. The middleware is responsible for establishing a link between the reader system and the users. The tags are attached to the goods or items that are intended to be identified.

The RFID transponders (tags) can be categorized based on the operating frequency such as (1) low frequency (LF: 120–150 kHz), (2) high frequency (HF: 13.56 MHz), (3) ultrahigh frequency (UHF: 866–868 MHz (European Union), 902–928 MHz (North American continent)), and (4) microwave

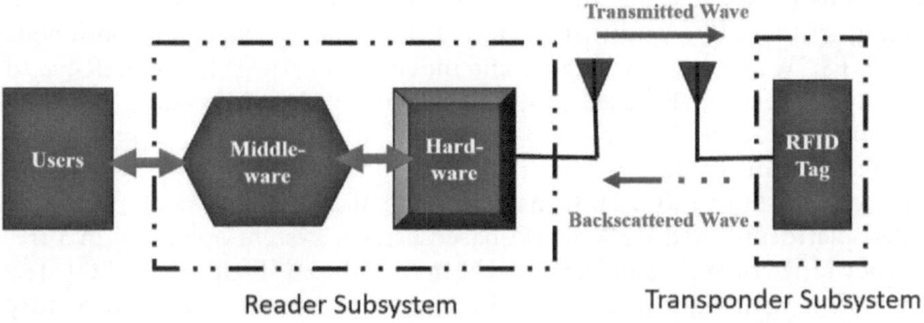

Figure 11.1 General RFID system.

bands (2.45 and 5.8 GHz). RFID is a non-line-of-sight radio-communication technology that helps to realize the IoT vision to create a global network of advanced systems [2]. All items could be identified and inventoried if they were fitted with radio tags. RFID transponders must be multi-bit featured, be low in cost, and consume less power to fulfill these functions. However, because of the higher cost of embedded integrated circuits and the power requirements, their widespread use is limited. As a result, cost-effective solutions are in high demand. RFID tags without silicon chips, called chipless tags, are now being produced for use in RFID systems. When several transponders are in the reader's field region, anticollision algorithms must be used to ensure proper tag recognition. A planar passive circuit is included in each chipless RFID tag. The reader will receive a unique EM signal from this planar circuit. This reflected signal is used to decipher the tag's identity. The majority of the tags use resonators based on microstrips. Jalaly and Robertson [4] were the first to report on a capacitively tuned dipole for RFID barcodes. Capacitively tuned identical microstrip dipoles constitute the tag antenna array designed at resonant frequencies. But capacitively tuned dipoles have unwanted parasitic effects and are limited in size. The number of resonators restricts the maximum number of bits that can be expressed by an RFID tag in the absence or presence of coding techniques. The tag's bit encoding capacity can be increased using the frequency shift coding (FSC) technique [1–3]. This approach is better suited to encoding large amounts of data with a small number of resonators, making it suitable for MIMO applications.

11.2.1 MIMO-equipped RFID for localization and identification

With advancements in the field of IoT, the identification and localization of goods or products are very much essential in smart warehouses [1]. RFID can be a suitable technique for both identification and localization of products in a vast range of applications such as logistics, inventory control, access control, automotive applications, and indoor localization [6–9]. Commercial off-the-shelf (COTS) interrogators/readers have been researched to a greater extent since the introduction of EPC Global UHF standards (Class-1, Generation-2) in the market [10–15]. The phase-evaluating systems like multi-frequency [12], frequency stepped continuous wave (FSCW) with a fixed reader and moving tags [16–19], and SAR-based [11] and inverse-SAR-based [20, 21] systems can provide better localization outcomes. In the case of 3-D product map generation applications, a huge amount of data is required for the identification and localization of goods in shops [20–22]. Considering the above requirements, a mobile robot platform-based UHF-RFID-based listener system operating in a frequency range of 865– 868 MHz has been introduced (Figure 11.2) [21]. The subsystem with eight antennas is highly useful in autonomous inventory control and localization of objects (in cm range) in the retail sector. The

Figure 11.2 UHF-RFID mobile robot listener subsystem [21].

design consists of a COTS reader with integrated MIMO antennas having some advanced features such as phase control of 360°, carrier cancellation in both radio frequency (RF) and intermediate frequency (IF) range, tag-decoder-based correlation, and position estimation.

The major challenge in RFID localization is accuracy. There are several techniques employed for RFID localization. SAR-based SISO RFID localization using a COTS interrogator is presented in [23]. If the aperture is greater, then the signal resolution is better. Similarly, a two-antenna SAR-based SISO method for tag localization is introduced in [13]. This means that a 2-D space tag location is accessed by an unmanned ground vehicle

(UGV). A COTS reader mobile robot environment is implemented to detect 37 transponders that are located on the ground [24].

Reader configurations having monostatic or bistatic characteristics are presented in [25–29]. In this literature, usage of the single-stage carrier cancellation or no carrier cancellation method is highlighted. In [21], a mobile robot platform with an eight-channel listener capable of 3-D position estimation is introduced. A significant increase in driving speed in comparison to monostatic systems is observed here. The architecture of the listener provides an increased resistance to multipath using multiple interrogators as a reading scheme, which is also useful to reduce the noise interference. With the additional features like automatic route planning, the system uses the simultaneous localization and mapping (SLAM) algorithm with SAR implementation. The system provides the position of the robot and does not require any additional vision-based systems [21].

11.2.2 System architecture: UHF-RFID listener

- The overall design of the UHF-RFID listener system (Figure 11.3) is comprised of three primary components: (1) the COTS UHF-RFID interrogator, (2) transponders, and (3) the listener subsystem [21].
- The Generation-2 (Gen-2) protocol (EPC Global) is partially employed in the listener subsystem architecture.
- This provides some advantageous features such as high flexibility, reduced complexity, and less computation.
- The reader unit has a transmitter (Tx) and a receiver (Rx) operating in monostatic mode. An unrestricted number of listeners on the Rx channels can be added to the system where the EPC Global UHF standard (Class-1 Gen-2) is followed by the listeners [10]. The reader signal is decoded to find out the communication parameters of the tag.
- The reader signal is given as input to a coupler (20 dB) trailed by a power divider (3 dB).The Wilkinson power divider is suitably used, and the output of the power divider is connected to (1) a reader signal distribution (RSD) board and (2) a clock-recovery and distribution (CRD) board.
- The outcomes of the RSD section are conditioned by amplifiers and attenuators. There are nine reader signals generated in this process, out of which eight are meant for RF carrier cancellation and one is sent to the baseband section.
- Similarly, the clock-recovery section is employed in the recovery of nine clock signals, out of which eight are passed to the RF board for amplification and attenuator control and one is transferred to the baseband board's mixer. At the mixer, it is transformed into a digital signal by the analog-to-digital converter (ADC) and fed to the full-programmable gate array (FPGA) unit.

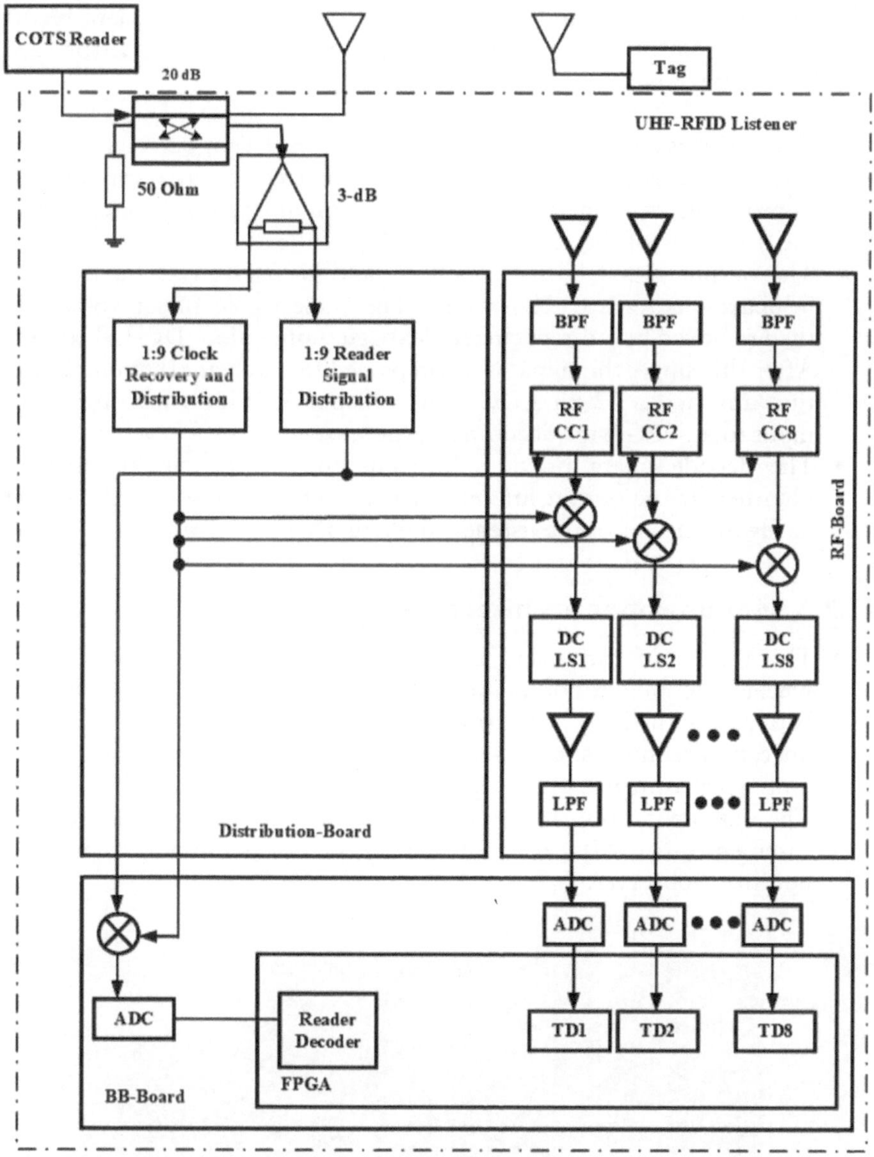

Figure 11.3 Listener system [21].

- Decoding of the tag signal is achieved by the envelope detector with the expression for the required backscatter link frequency (f_{BLF}) given in Equation (11.1) [21]:

$$f_{BLF} = \frac{DR}{T_{TRcal}} \tag{11.1}$$

where DR = divide ratio and T_{TRcal} = time period between the data bits 0 and 1. Extraction of both the parameters can be accomplished from the signal of the interrogator unit.

- The major issue in UHF-RFID is the poor isolation between Tx and Rx. Bistatic MIMO may have strong interference at reception. To improve the performance, carrier suppression circuitry is included at the first stage of the receiving path [21].
- The output signal is first amplified and then given as input to an in-phase-quadrature (IQ) mixer. The undesirable offset voltage is then removed by a direct current leakage suppression (DC-LS) circuit. After this stage, the signal is again passed through another amplification unit and a 12-bit ADC. The transponder's information is then input to an FPGA for decoding purposes.
- The decoded data provide the signal phase and signal intensity obtained at the output for further use. Figure 11.4 shows the whole hardware circuit of the listener subsystem.

11.2.3 Mobile robot architecture

- The automatic movement of antennas is archived by a differential wheeled mobile robot platform powered by rechargeable lithium batteries with the specification 28.8V DC/40 Ah. The mobile robot operation requires some essential sensors such as the motor encoding unit, accelerometer, gyroscope, and distance sensors based on lasers. The outcomes from these sensors are combined in an internal measurement unit (IMU) to create a state vector estimation by using a nonlinear observer.

Figure 11.4 The listener hardware [21].

- The state vector has some specific features such as x, y coordinates, angular orientation Θ, and velocity information. Autonomous navigation is enabled by generating a two-dimensional location map of the platform. The occupancy map is generated by analyzing the laser scanner data and state vector information.
- The SLAM algorithm is employed for the system requirement analysis.
- The estimation of position is obtained by using the occupancy map, the range information from the laser scanning unit, and the dynamic model of the mobile robot with the SAR platform.

11.2.4 Antenna and system theory

- The system configuration and the antenna placement are very much essential for estimating position accurately. The simulated beam widths at a transmission frequency of 866.3 MHz for MOMO, SISO, and SISO 2 setups are analyzed in Figure 11.5. The multi-antenna environment of eight antennas placed in two different rows can be used for MIMO interlacing (Figure 11.5) [21]. This approach provides a unique localization result with a distance among antenna elements less than $\lambda_c/4$ m.
- SISO interlacing, with antennas placed closer than $\lambda_c/4$ m to each other, yields a higher side lobe level. There is no special position estimate in the setup of the SISO 2. As the distance between two readers is 1.07 m, the z-direction spatial-sampling theorem does not match. The output power of the beamformer is analyzed for a fixed value z and a varying value x with the same configuration. The value of z is considered as 0.2 m.

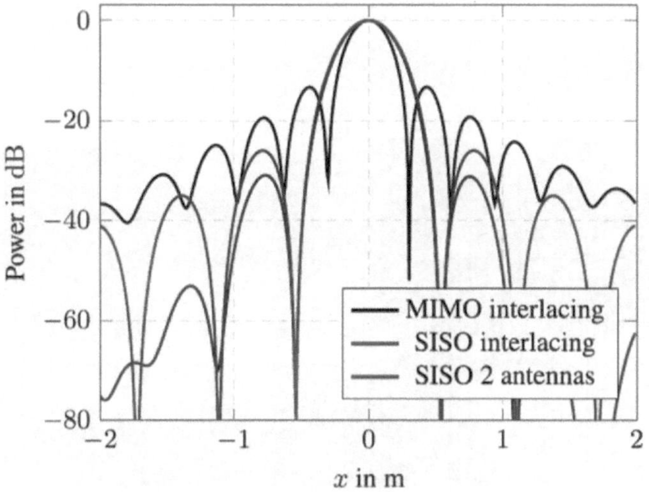

Figure 11.5 Beamformer output power [21].

11.2.5 Measurement results

Position estimation is performed by taking ten tags that require the z-direction spacing between two antennas to be less than $\lambda_c/4$ m (by incorporating multiple rows). The antenna has the property of circular polarization with a gain of 7.5 dBi. It also characterized by a beam width of 70°. The obtained accuracy of 1.45 cm can be considered a standard for industrial implementation. Then the number of tags is increased to 214 to evaluate system performance. The outcome of position estimation using the 214 transponders is presented in [21], indicating the tag and antenna positions. Different color codes and symbols are used to present vital information about system estimation. The system can be useful in inventory control, stock management, and smart warehouses for IoT applications. A real warehouse environment with a mobile robot platform is shown in Figure 11.6. The platform is capable of detecting all the possible obstacles by using laser scanners, and from the scanned data, a map of these can be generated. This vital information helps to orient the robot on the map and provides localization information with good accuracy (1 cm). This kind of system is highly capable of reducing manpower and costs in the industrial sector.

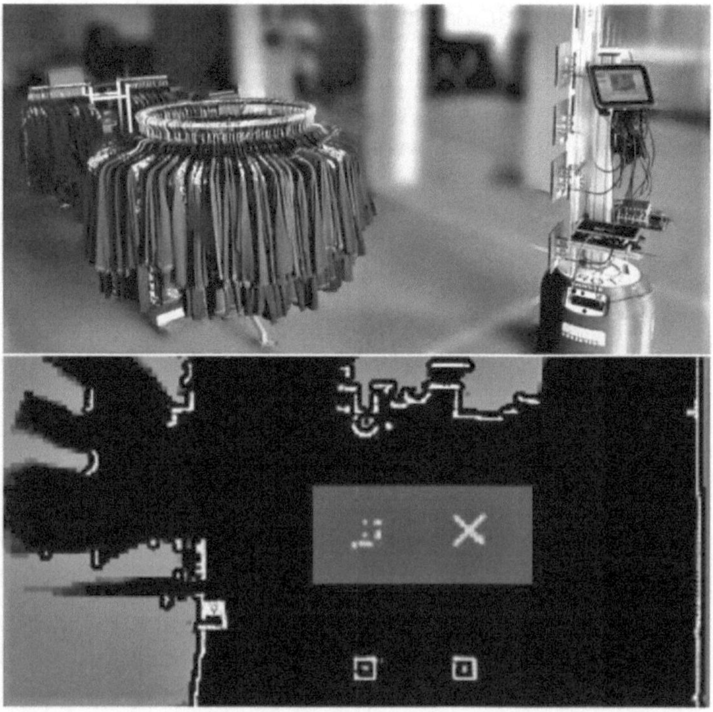

Figure 11.6 RFID mobile robot equipped warehouse [21].

11.3 MIMO APPLICATION IN RFID

11.3.1 Backscatter channel measurements

The channel information must be used to increase the efficiency of the backscatter connection or to increase the data rate by using MIMO techniques. The approach is based on the principle that a multiport (n-port) with a reduced number of measuring ports will disperse parameters. In this case, load impedances are usually used for transmitting information through backscatter modulation within the RFID tag. The technique could be used for the future generations of the multi-antenna RFID tags by evaluating the capacity of the channel for the MIMO-RFID prototype systems. For measurement purposes, two readers for transmitting, two readers for receiving, and two tag antennas (2 × 2 × 2) are considered at a frequency ranging from 5 to 6 GHz. Because of the need for higher data rates, range, and reliability, the main focus of research is multi-antenna RFID systems with higher operating frequencies that allow the MIMO technique to be applied. Multiple antennas can also be used inside RFID systems in the reader and on the transponder. A general M × L × N MIMO-RFID system is depicted in Figure 11.7. In the narrowband case, the system can be characterized with the complex baseband channel coefficients h^f_{xy} and h^b_{xy} for the forward and the backscatter links, respectively. With the Rx series (N > 1), to improve the efficiency or range by maximum ratio combining (MRC) or to retrieve many tags from a collision, MIMO can be applied. A Tx array (M > 1) can be used to increase the backscatter power of the transponder [30]. Using rectangular-shaped resonators, the multibit coding capacity of the passive transponders can be increased [31, 32].

The signals matrix will transform into a normalized identity matrix if the signal is modified by the same antenna and no internal transmission exists between the transmitters. The signals matrix is a diagonal one if

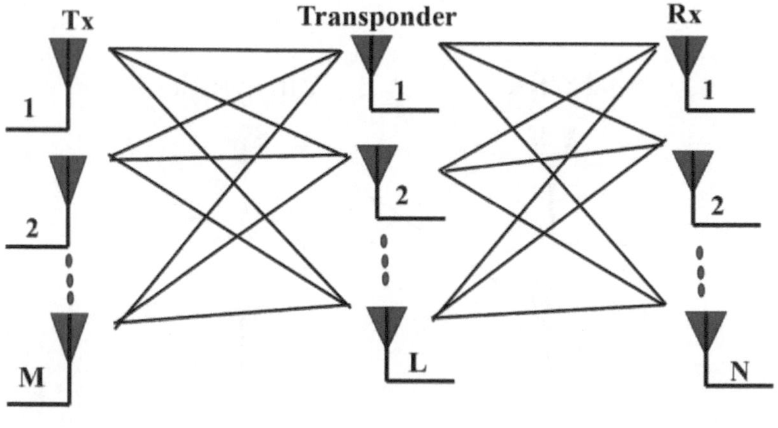

Figure 11.7 RFID-based MIMO system (M × L × N).

each antenna is modulated by individual data streams. This type can also be used to improve data rates with the SM technique or to increase reliability with space-time coding. The entire signals matrix can be obtained if further transmission between the tag antennas is present. To date, few implementations have been seen for this kind of signal matrix—e.g., where non-diagonal elements are used to create retro directive arrays. Channel knowledge is the most important parameter when implementing MIMO techniques in RFID systems. The channel information of the forward and reverse channels is required for the optimum use of the MIMO, considering dynamic channel measurements [30].

Here, the forward link and the backward link are identical, so the general scheme of the $M \times L \times N$ system design (Figure 11.7) is reduced to Figure 11.8. The same antenna of the reader subsystem can serve as both transmitter and receiver, thereby reducing the extra circuitry and cost. The considered channel coefficients are replaced by the scattering parameters ($S_{13} = S_{31}$, $S_{14} = S_{41}$, $S_{23} = S_{32}$, and $S_{24} = S_{42}$), which are highlighted in yellow in Figure 11.9. S_{11}, $S_{12} = S_{21}$, and S_{22} (green) are taken into account, representing the reflection and transmission parameters, which are responsible for measuring backscatter modulation. As chipless RFID transponders do not have absolute recipient hardware circuits on the board, the channel parameters can be obtained by measuring the reader ports. It is therefore important to have a method for determining the maximum dispersion matrix without calculating all the port extraction parameters. For this, a port reduction method (PRM) [33, 34] based on the above theory (renormalization of a scattering matrix) [35] is applied. The Type I PRM is calculated unequivocally in S_{11}, $S_{12} = S_{21}$, S_{22}, S_{33}, and S_{44} (green in Figure 11.9). However, since the backscatter relation parameters (yellow in Figure 11.9) for the reciprocal n-ports (such as the multiple-interrogator RFID channel) are found by the calculating the square root of the squared parameter, there are independent signs (180° phase) for each parameter. A further calculation linking the

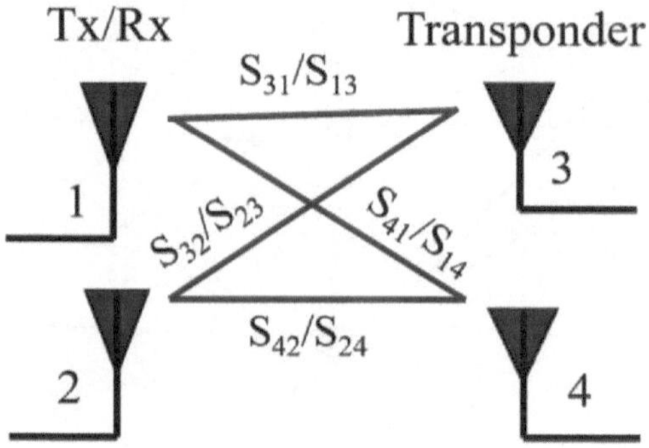

Figure 11.8 MIMO-RFID system (2 × 2 × 2).

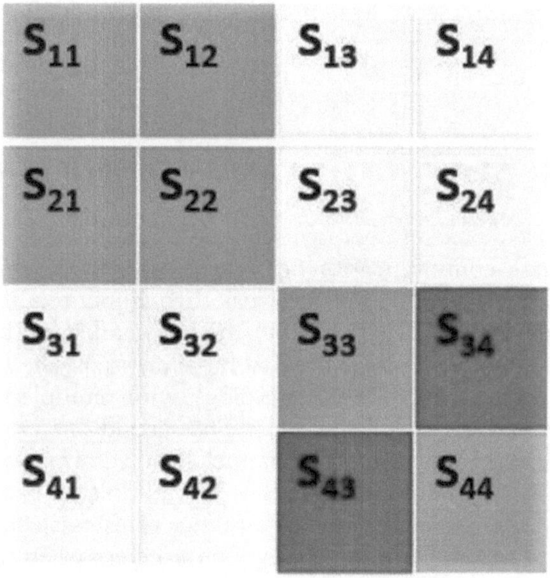

Figure 11.9 Four-port S-parameters for 2 × 2 × 2 MIMO-RFID.

vector network analyzer (VNA) to port n will overcome this [34]. Because this is impractical for true tags, another way must be sought to remove the uncertainty. Thus, an additional subset calculation between tag antennas is suggested with an internal through norm Γ_T. The elements apart from the main diagonal of the tag signals matrix are used to connect various parameters for the backscattering relation. $S_{34} = S_{43}$ (red in Figure 11.9) reflects the near-field connection and the far-field reflection of both tag antennas. The magnitude of far-field radiations is degraded by the path loss, so near-field coupling is expected to lead S_{34} [30].

11.3.2 Results and analysis

By taking 2 × 2 × 2 setups containing two interrogators and two transponder antennas, the measurements are conducted. By taking two self-complementary readers attached to two Agilent SP6T switches (model-L7106B-T24) serving as adjustable loads, the tags of two semi-passive antennas are emulated (an Agilent calibration package is used). Although not specifically needed, a supplemental calculation is carried out with scattered tag antennas to reduce uncertainty further. Backscatter channel coefficients in magnitude and phase are presented to detect the tag position. Compared with the reference calculation, the reduction of sign ambiguities is successfully verified with two solutions that are common to all channel coefficients. The magnitude and the phase are very similar for the backscatter coefficient and reference measurements. System measurement results for a device with two interrogators and two transponders are provided, and their agreement with a reference measurement [30] validates the procedure.

11.4 MIMO IN WLAN WITH HYBRID BEAMFORMING

With the increase in the number of users and the number of antennas that are used in a single communication device, the capacity of wireless communication network also grows. This network throughput requirement can be fulfilled by large-scale MU-MIMO. But the key challenges that hinder the advancement of this technology need to be addressed. The first challenge is that today's expansive MIMO technology with multiple users requires a large number of high-cost RF chains. The second one is that the wireless AP will be overloaded due to channel state information (CSI) feedback intended to eliminate interference from multiple users and antennas. The third challenge is that the lack of an efficient user-selection algorithm limits the ability to calculate the channel parameters when the number of users are larger. To solve these problems, BUSH (a large-scale MU-MIMO solution prototype) has been developed. This model uses hybrid digital-analog beamforming to make a scalable selection of user beams for older Wi-Fi phased-array antennas. BUSH introduces a low-cost testing scheme on multicarrier WLANs and develops a high-precision power azimuth spectrum (PAS) assessment algorithm for one RF chain only. The antennas use analog beamforming to direct the beam to each chosen downlink user to minimize the number of RF chains. Using the beamforming technique, further reduction of interference can be achieved. BUSH is deployed on a software-defined radio (SDR) network and has proved its success in over 30 different indoor setups. The model of the WLAN has recently been updated with the adoption of the IEEE 802.11ac standards from single users to multiple users [35–36].

We are now seeing more and more wearable Wi-Fi gadgets, adding larger data transfer loads to existing Wi-Fi facilities. The number of personal mobile appliances attached to our bodies is increasing gradually. Although 802.11ac is commonly used in universities and businesses to minimize the growing connectivity issues, other difficult issues like the selection of users persist unresolved. 802.11ac uses MU-MIMO, which enables an access point to send separate data streams to different devices in a single frequency range by multiple antennas at one time. The user tests the channel to the AP and sends the information about the channel condition to the AP. Then a user receives only individual data streams, with any unwanted (or interfering) data streams excluded. Zero forcing beam formation (ZFBF) is a popular beamforming method that allows users to remove redundant data sources. This enables the MU-MIMO AP to use the antenna resources better and achieve a good performance increase. Theoretically, with the rise in the number of antennas, the downlink performance increases linearly. The new practice is to introduce a large antenna array on the AP side so that the MIMO WLAN potential is maximized. The MU-MIMO-wide network capacity is being further enhanced by connecting more than ten antennas to the transmitter [36–39]. In MU-MIMO, the transmitter's baseband is beamformed, which involves the use of a single RF chain for each antenna. The whole RF circuit includes several components such as the baseband processor, modulator-demodulator, and digital-to-analog/analog-to-digital converter (DAC/ADC) [36], among other things.

11.4.1 Challenges and solutions in MIMO implementation

Although MU-MIMO has successfully carried out large-scale prototyping in mobile communication systems, the following challenges are still faced in actual deployment:

- First, the cost of the RF chain is more for the large-scale MU-MIMO device to link any antenna to the RF chain.
- Second, with the rise in antenna numbers, the overhead increases linearly. However, the CSI received sequentially from all users can cause a considerable delay and consequently worsen the results.
- Third, in the IEEE 802.11ac standard, the user selection strategy is ambiguous but plays a major role in the transfer operations. The num ber of interrogators (antennas) linked to the base system in 802.11ac is restricted to eight with more users. The AP must then pick a community of users in every slot to obtain the time slot.

11.4.2 Overcoming challenges

In the following, the design, execution, and assessment of BUSH and a large-scale, 802.11ac-compliant MU-MIMO system are discussed. Using hybrid beam shaping, BUSH performs a combination of beamforming techniques and a user selection algorithm. The theory is chosen to minimize cross talk by selecting an individual user (or multiple users) with different angle of arrival (AoA) peaks in the selected category. The MUSIC (MUltiple SIgnal Classification) AoA is a continuum of probabilities but not a power function. Therefore, it is proposed to derive the efficient power in each direction of the AoA by a new blind estimation system with the obtained power. There are two benefits of BUSH. Using the MIMO technique, multiple antenna components can be used to obtain a high DBF gain and the multiplexing gain. The performance efficiency does not decline because the RF chain is distributed to all users. On the other side, scalable beam consumer connectivity can effectively minimize cross talk in analog beam formation where the number of customers is more than the number of RF chains. RF chains can also be used to get the required feedback from the receiving units to minimize the channel interference between various receivers. The actual application of BUSH in WLAN brings many advantages and some challenges. In 802.11 WLAN, it is not easy to calculate the AoA of multiple users at the same time. The Wi-Fi system is not centrally controlled, so, unlike with the Long-Term Evolution (LTE) system, the user does not always send the pilot to the AP. As the sounding procedure in the Wi-Fi system is longer, the choice of user beam becomes a combined decision-making problem. Therefore, the focus should be on the following points:

- The design of BUSH, the digital-analog HBF MIMO system, enables the system to eliminate the mismatch between a limited RF chain and a large number of antennas.
- Using different subcarriers for different users, AoA information can be received from multiple users at the same time within a single round of probing, thereby greatly reducing the delay.
- Because the traditional MUSIC algorithm does not work satisfactorily for multiple antennas, blind AoA estimation is proposed, which obtains accurate AoA information by using the PAS method to evaluate power expression.
- An optimized framework is designed for joint beamforming and user selection to achieve better performance. In addition, trace-driven emulations of BUSH are carried out for file transmission, video transmission, and adaptive-dynamic streaming via HTTP [36, 39].

Further, the following aspects can be considered to overcome the issues listed in Section 11.4.1:

- An HBF architecture that uses multiple frequency bands to control a wide range of phased-array antennas can be the solution to the first problem. It is a two-layer beamforming architecture and includes the architecture of analog beamforming (ABF), which uses a phase shifter to monitor the signal phase of each antenna. ABF can provide beamforming benefits for each RF channel, and DBF can eliminate reciprocal interaction among users to achieve a spatial increase in multiplexing. It should be remembered that the exact number of RF chains includes pure DBF, which is often impractical.
- The other two problems can be solved by approaching BUSH technique. This solution gives information about the AoA, the power requirement status of all users, and the received information in order to select the individual users. Therefore, the channel input information from the chosen customers is the only requirement, which reduces the overhead and latency considerably.

11.4.3 System evaluation

Initially, the circumstance is calculated by taking two users with two RF chains. A more flexible beam and user selection algorithm is developed for BUSH compared with Hekaton [40]. The method is not an individual mapping solution, such as that used by Hekaton. This implies that if two beams that support a user are higher than two separate users, two beams will serve a single user. It can be verified that the operating performance of the system improves by 8 percent on average compared to Hekaton. BUSH includes all user information for optimum overall performance in the transmission with low signal-to-noise ratios (SNRs), whereas Hekaton can support two users in all situations. BUSH, therefore, is particularly suited for use with extremely low SNRs in multiuser situations [40]. Next, to check multiple-user scalability, the total user count is raised from 2 to 10, and the total output of the network is recorded. For multiple users, BUSH still exceeds Hekaton and Vanilla 802.11ac. Furthermore, from the cumulative distribution function (CDF), the average improvement for BUSH over Vanilla 802.11ac is 108.6 % and over other counter parts is 22 %. These enhancements come largely from the selection of the HBF joint consumer beam [36].

11.4.4 Different scenarios

BUSH has been tested in various indoor conditions to show its reliability. A meeting room of 5.5–6 m^2 and an office of 10–12 m^2 were selected for the evaluation of the CDF and BUSH throughput plots. BUSH throughput surpassed that of Hekaton and Vanilla 802.11ac by 28.55 percent in the meeting room and 160.10 percent on average. The overall throughput increased by 26.87 percent in the office and 133.56 percent on average relative to Hekaton and Vanilla 802.11ac. It has been observed that the BUSH gains are higher in performance evaluation. The explanation is that the office and meeting room sizes are relatively smaller than the classroom. Thus, the number of users is greater, so more people can choose from a user-selection algorithm in the same location. In summary, the BUSH is strong against structural changes.

11.4.5 Energy efficiency

Energy efficiency (bits/Joule) is compared among traditional 802.11ac MU-MIMO, Hekaton, and BUSH. Higher energy efficiency is achieved by implementing BUSH because its power demand is small and it uses a staggering range of energy efficiency improvements. BUSH is found to be marginally higher than Hekaton in its performance. The explanation for this is that using two RF chains to represent one consumer is one of the suitable techniques in some low-scale SNR situations, but Hekaton fails to do this. The combined user-beam selection with hybrid beamforming is largely responsible for these advances.

11.4.6 Some related applications

11.4.6.1 ABF

A codebook is designed for ABF in millimeter wave (mmWave) with a general requirement that it needs to search all the codebook entries. As the number of antenna elements increases, the coordination overhead increases. For ABF in MIMO systems, a phase shifter is adopted for downlink scheduling in the MIMO-WiMAX network [37, 39].

11.4.6.2 HBF

Both ABF and DBF beamforming techniques are used here. ABF uses phase shifters, while DBF uses RF chains for channel measurements. HBF is designed for single and multiple users based on compressive sensing in mmWave communications [39, 40]. The existing work focuses on the theoretical concept regarding fully DBF and ABF where the numbers of RF chains and data streams vary. For existing LTE systems, new HBF architecture has been designed (i.e., Hekaton [40]). Hekaton has implemented a compression-based AoA estimation algorithm and a downlink selection metric to permit large-scale MIMO [36].

11.5 FABRIC BODY-WORN MIMO SYSTEM

Recently, considerable research has been conducted in the field of flexible electronics in reaction mainly to the rising demand for lightweight and portable smartphones [41]. Making use of knowledge of the additive manufacturing processes and the suitable substrate is essential for the affordable mass manufacture of these flexible electronic printed items such as RFID tags, wireless sensors, conduction tags and printers, keypads, and display applications. Cellulose-based substrates (CBSs), such as papers and plastics, are reusable and versatile and can be printed directly with conductive inks. Different printing methods, such as flexographic (flexo), gravure, slide, inkjet, aerosol, and fine film, use organic material printed in conductive inks [2]. For wearable devices, the wearable patch antenna has a semi-directional radiation pattern, which is preferred over the omnidirectional radiation pattern of a traditional dipole antenna to prevent unnatural human body radiation interference and lack of radiation. Regarding antenna correlations and efficient gain in various wireless system models, the special effects of the antenna direction and position on MIMO device performance are required to evaluate system performance. Complex physical modeling by combining ray-tracing full-wave EM simulations has confirmed the superiority of a wearable device over a traditional system. The channel measurement outcomes for a body-worn antenna system reveal a considerable increase in system capacity compared to a reference full-size dipole antenna [41].

11.5.1 Distributed wearable MIMO system

The increased need in most wireless networks for higher data rates and greater system output has been the force driving the technical transition in wireless communications. MIMO, the current advanced antenna technology, can improve both specific performance and connection robustness and has already been applied to many wireless standards (e.g., IEEE 802.11n and 802.16e). MIMO strategies use multipath reflections to create several simulated spatial networks by numerically unwrapping the wireless signal paths rather than tackling multipath fading issues found in most wireless settings. A distributed wearable MIMO device using electro-textile directional antennas is discussed in [41]. This innovative wearable antenna will significantly increase MIMO device capability, when contrasted with the limited sub-wave duration experienced by other technologies, by positioning antennas in various positions on human clothing. In an internal office setting, major diversity gains have been found with a wearable multi-antenna device. In the literature [41–45], detailed research is described on the use of omnidirectional readers (dipole antennas) in multiple-antenna-based systems. The outputs of MIMO-based wireless systems that consist of both patch antenna arrays and sector antennas arrays have been analyzed. Through theoretical derivation and numeric simulation, it is possible to maximize antenna diversity by using a wearable directional antenna array. MIMO capability can be superior when compared to traditional solutions like the uniform linear array (ULA) of antennas. More elaborate studies in typical wireless environments have also been done, showing the potential significant advantages of the MIMO-based body-worn antenna system. Taking into account different wearable-specific parameters and challenges such as the effects of close human contact, slight wear from the bending of the fabric antenna, signal correlation, mutual coupling, and power imbalance, the system has been evaluated. To achieve the required system performance through a multiple-transceiver body-worn device, a 3-D model has been designed and human proximity and radiation properties studied [41].

11.5.2 Body-area antenna design

Complete wave and analytical modeling has been conducted to analyze the impact of closeness on wearable antennas' radiation properties [2, 41]. The simulation program is employed in the Ansoft HFS Simulator (HFSS). This interrogator consists of a perfect electricity-conducting ground layer, a metal patch sheet, and an insulation layer. A coaxial attachment to the ground plane supplies the patch antenna. A truncated three-layer human design comprises a skin layer (1 mm wide), a fat layer (3 mm wide), and a muscular layer (40 mm wide). This is located 2 mm from the antenna to investigate the local interaction of the EM fields. The measured radiation antenna patterns with human proximity remain very close to the virtual antenna patterns, particularly on the main beam, because of the inclusion of a metal plate as the ground in the patch antenna configuration. The presence of the protective layer restricts contact with the muscle layer of the body. The middle antenna operating frequency shifted by approximately 1 percent, and due to human proximity, the antenna effectiveness decreased by approximately 10 percent. These effects of simulation correspond to related studies, but the real analysis of human models is more complicated. When changing the spacing between the antenna and the tissue between 2 and 4 mm, the radiative property of the antenna is indifferent to the distance of the separation. Furthermore, the theoretical radiation model is closely consistent in free space with the virtual antenna pattern. The smaller theoretical back radiation is due to the presumption of a scientific derivation of an infinite ground plane. The infinite ground plane assumption is suitable to analyze the efficiency for wearable MIMO-based patch antennas (by speeding up the solution to find out the antenna parameters).

11.5.3 Antenna array for wearable fabric

Directional patch antennas can be used in ULAs for fabric body-worn elements. At first setup, the fabric body-worn antennas are placed on the garment (front and back) [41]. In this case, only a broadside orientation is obtained by using an array of multiple antennas. This array of antennas uses only the space diversity technique. The antennas are placed on all sides of the garment in the second setup. Here, directional antennas cover all large antenna panel directions. The collection of antennas incorporates the diversity of space and patterns. This arrangement explores pattern and spatial diversity in the MIMO communication framework. For both transmitting and receiving antennas, vertical polarization is considered. However, the plurality of polarizations can be seen along the same lines as with a traditional medium (for example, a mobile device). In systems with multiple antenna arrays, a dual-polarized patch array can produce channels that are orthogonal to each other during indoor channel measurements. The same is true for all antennas that are orthogonally polarized, regardless of the original location. Thus, this is not a problem specific to fabric on-body antennas. At the corner of the hall outside the test space, four interrogators (dipole antennas) were placed with 5 cm of space between neighboring readers. One individual holding the receiver antenna roamed randomly in a small office with a standard interior area (3.2—3.7 m^2). There were four absolute fabric patch antennas on the reception antenna array. Two patch antennas were placed on the front of the fabric cost, and the remaining two were placed on the back. The patch antenna parameters used for the fabric are calculated and the details are discussed in [41]. In addition, during the measurement, a reference half-wavelength dipole antenna was considered with a 2.2 dBi realized gain at 2.4 GHz. A VNA, along with two electromechanical switches, was used to test the vector chain reaction between the transmitting and the receiving antennas attached to the VNA output port and the receiving antenna attached to the VNA input port. A 50 Ω load was applied to all the unselected antennas. For excellent and reliable efficiency, the electromechanical switches were chosen with maximum insertion loss and minimum isolation of 0.2 dB and 80 dB, respectively. The channel response time between each transmission and the reception of antennas was 0.5 s, including 15 ms for switching, and 25 ms of VNA measuring time, with other delay times considered for measurement calculations. The complex matrices for the body-worn patches and the dipole antennas were calculated, and the MIMO system power was statistically analyzed. At SNR = 10 dB on a single dipole antenna, the CDFs with the MIMO power were traced ($N_r = N_t = 4$). On the single dipole antenna, the outage potential was 1.6 b/s/Hz, and on the two-sided body-worn patch array it was 6.6 b/s/Hz. The outage capacity of the planned two-sided patch array exceeded that of the single antenna by 319 percent [41].

11.6 SUMMARY AND CONCLUSIONS

- The chapter summarizes a wide range of applications of MIMO-based systems, including RFID localization, wide area networks, and IoT.
- There can be further advancements in the fields of RFID localization and mobile robot technology in MIMO systems. A brief discussion on MIMO-based RFID systems for channel measurement methods has been presented. The technique is accomplished with multiple readers and tag antennas. This method analyzes the system in terms of scattering parameters and also characterizes the channel by evaluating the performance with MIMO capacity.
- The challenges in implementing the MIMO system and the procedures for overcoming them are discussed to enable advancements in localization, tracking, and identification of objects.
- Directivity and location of antennas in the MIMO system have been analyzed, and the parameters such as gain, antenna correlation, and energy efficiency have been evaluated with various wireless channel models.
- A comparative study between the conventional system and the wearable system that shows the advantages of the latter has been presented with the help of full-wave simulation. Along with this, the different MIMO beamforming techniques (analog, digital, and hybrid) have been discussed for a vast range of applications.

REFERENCES

[1] K. Finkenzeller, *RFID Handbook: Radio-Frequency Identification Fundamentals and Applications*. Hoboken, NJ: Wiley, 1999.
[2] S. K. Behera and N. C. Karmakar, "Chipless RFID printing technologies: A state of the art," *IEEE Microwave Mag.*, vol. 22, no. 6, pp. 64–81, June 2021, doi: 10.1109/MMM.2021.3064099.
[3] N. C. Karmakar, ed., *Handbook of Smart Antennas for RFID Systems*. Hoboken, NJ: Wiley, 2011.
[4] I. Jalaly and I. Robertson, "RF barcodes using multiple frequency bands," *IEEE MTT-S Digest*, vol. 4, pp. 139–142, 2005.
[5] A. Ghelichi and A. Abdelgawad, "A study on RFID-based kanban system in inventory management," *Proc. IEEE Int. Conf. Ind. Eng. Eng. Manage.*, Bandar Sunway, Malaysia, Dec. 2014, pp. 1357–1361.
[6] E. Hidalgo, F. Munoz, A. Guerrero de Mier, R. G. Carvajal, and R. Martin-Clemente, "Wireless inventory of traffic signs based on passive RFID technology," *Proc. 39th Annu. Conf. IEEE Ind. Electron. Soc.*, Vienna, Austria, Nov. 2013, pp. 5467–5471.
[7] Y. Liang, J. Xiao, Y. Wang, Z. Pang, Y. Li, X. Jin, and G. Liu, "A method to make accurate inventory of smart meters in multi-tags group-reading environment," *Proc. IEEE Int. Conf. RFID Technol. Appl.*, Foshan, China, Sep. 2016, pp. 123–128.
[8] M. Hehn, E. Sippel, C. Carlowitz, and M. Vossiek, "High-accuracy localization and calibration for 5-dof indoor magnetic positioning systems," *IEEE Trans. Instrum. Meas.*, vol. 68, no. 10, pp. 4135–4145, Oct. 2019.
[9] V. Stanford, "Pervasive computing goes the last hundred feet with RFID systems," *IEEE Pervas. Comput.*, vol. 2, no. 2, pp. 9–14, Apr. 2003.
[10] *Class-1 Generation-2 UHF RFID Protocol for Communications at 860 MHz–960 MHz, Version 1.2.0.*, EPCglobal, Brussels, Belgium, Oct. 2008. https://www.gs1.org/sites/default/files/docs/epc/Gen2_Protocol_Standard.pdf.
[11] A. Parr, R. Miesen, and M. Vossiek, "Comparison of phase-based 3D near-field source localization techniques for UHF RFID," *Sensors*, vol. 16, no. 7, p. 978, Jun. 2016.
[12] C. Li, L. Mo, and X. Xie, "Localization of passive UHF RFID tags on assembly line based on phase difference," *Proc. IEEE Int. Instrum. Meas. Technol. Conf. Process.*, Taipei, Taiwan, May 2016, pp. 1–6.
[13] A. Motroni, P. Nepa, P. Tripicchio, and M. Unetti, "A multi-antenna SAR based method for UHF RFID tag localization via UGV," *Proc. IEEE Int. Conf. RFID Technol. Appl.*, Macau, China, Sep. 2018, pp. 1–6.
[14] R. Miesen, A. Parr, J. Schleu, and M. Vossiek, "360° carrier phase measurement for UHF RFID local positioning," *Proc. IEEE Int. Conf. RFID Technol. Appl.*, Johor Bahru, Malaysia, Sep. 2013, pp. 1–6.
[15] X. Li, Y. Zhang, and M. G. Amin, "Multifrequency-based range estimation of RFID tags," *Proc. IEEE Int. Conf.*, Orlando, FL, Apr. 2009, pp. 147–154.
[16] M. Scherhäufl, M. Pichler, D. Muller, A. Ziroff, and A. Stelzer, "Phase-of--arrival-based localization of passive UHF RFID tags," *IEEE MTT-S Int. Microw. Symp. Dig.*, Seattle, WA, June 2013, pp. 1–3.

[17] M. Scherhäufl, M. Pichler, E. Schimback, D. J. Muller, A. Ziroff, and A. Stelzer, "Indoor localization of passive UHF RFID tags based on phase-of-arrival evaluation," *IEEE Trans. Microw. Theory Techn.*, vol. 61, no. 12, pp. 4724–4729, Dec. 2013.
[18] M. Wegener, D. Fros, M. Rosler, C. Drechsler, C. Patz, and U. Heinkel, "Relative localisation of passive UHF-tags by phase tracking," *Proc. 13th Int. Multi-Conf. Syst., Signals Devices*, Leipzig, Germany, March 2016, pp. 503–506.
[19] L. Yang, Y. Chen, X.-Y. Li, C. Xiao, M. Li, and Y. Liu, "Tagoram: Realtime tracking of mobile RFID tags to high precision using COTS devices," *Proc. 20th Annu. Int. Conf. Mobile Comput. Netw.*, Maui, HI, 2014, pp. 237–248.
[20] A. Parr, R. Miesen, and M. Vossiek, "Inverse SAR approach for localization of moving RFID tags," *Proc. IEEE Int. Conf. RFID*, Penang, Malaysia, April 2013, pp. 104–109.
[21] M. Gareis et al., "Novel UHF-RFID listener hardware architecture and system concept for a mobile robot based MIMO SAR RFID localization," *IEEE Access*, vol. 9, pp. 497–510, 2021, doi: 10.1109/ACCESS.2020.3047122.
[22] C. Johnston, *Amazon Opens a Supermarket with No Checkouts*, 2018 [Online]. Available: https://www.bbc.com/news/business-42769096.
[23] A. Buffi, M. R. Pino, and P. Nepa, "Experimental validation of a SAR based RFID localization technique exploiting an automated handling system," *IEEE Antennas Wireless Propag. Lett.*, vol. 16, pp. 2795–2798, 2017.
[24] A. Motroni, P. Nepa, V. Magnago, A. Buffi, B. Tellini, D. Fontanelli, and D. Macii, "SAR-based indoor localization of UHF-RFID tags via mobile robot," *Proc. Int. Conf. Indoor Positioning Indoor Navigat.*, Nantes, France, September 2018, pp. 1–8.
[25] C. Angerer, "A digital receiver architecture for RFID readers," *Proc. Int. Symp. Ind. Embedded Syst.*, Le Grande Motte, France, June 2008, pp. 89–94.
[26] R. Langwieser, G. Lasser, C. Angerer, M. Rupp, and A. L. Scholtz, "A modular UHF reader frontend for a flexible RFID testbed," *Proc. 2nd Int. Workshop RFID Technol.*, Vienna, Austria, July 2008, pp. 1–5.
[27] P. Nikitin, S. Ramamurthy, and R. Martinez, "Simple low cost UHF RFID reader," *Proc. IEEE Int. Conf. RFID*, Orlando, FL, April 2013, pp. 126–127.
[28] P. Pursula, M. Kiviranta, and H. Seppa, "UHF RFID reader with reflected power canceller," *IEEE Microw. Wireless Compon. Lett.*, vol. 19, no. 1, pp. 48–50, January 2009.
[29] N. Decarli, "On phase-based localization with narrowband backscatter signals," *EURASIP J. Adv. Signal Process.*, vol. 2018, no. 1, p. 70, Dec. 2018.
[30] E. Denicke, M. Henning, H. Rabe, and B. Geck, "The application of multiport theory for MIMO RFID backscatter channel measurements," *2012 42nd Eur. Microwave Conf.*, Amsterdam, Netherlands, 2012, pp. 522–525, doi: 10.23919/EuMC.2012.6459123.
[31] D. P. Mishra, T. Kumar Das, P. Sethy, and S. K. Behera, "Design of a multi-bit chipless RFID tag using square split-ring resonators," *2019 IEEE Indian Conference on Antennas and Propagation*, Ahmedabad, India, 2019, pp. 1–4, doi: 10.1109/InCAP47789.2019.9134626.

[32] D. P. Mishra, T. K. Das, and S. K. Behera, "Design of a 3-bitchipless RFID tag using circular split-ring resonators for retail and healthcare applications," *2020 Nat. Conf. Commun.*, Kharagpur, India, 2020, pp. 1–4, doi: 10.1109/NCC48643.2020.9056018.
[33] J. D. Griffin, and G. D. Durgin, "Multipath fading measurements at 5.8 GHz for backscatter tags with multiple antennas," *IEEE Trans. Antennas Propag.*, vol. 58, no. 11, pp. 3693–3700, Nov. 2010.
[34] H. Lu and T. Chu, "Port reduction methods for scattering matrix measurement of an n-port network," *IEEE Trans. Microwave Theory Tech.*, vol. 48, no. 6, pp. 959–968, 2000.
[35] J. Tippet and R. Speciale, "A rigorous technique for measuring the scattering matrix of a multiport device with a 2-port network analyzer," *IEEE Trans. Microwave Theory Tech.*, vol. 30, no. 5, pp. 661–666, May 1982.
[36] Z. Chen, X. Zhang, S. Wang, Y. Xu, J. Xiong, and X. Wang, "Enabling practical large-scale MIMO in WLANs with hybrid beamforming," *IEEE/ACM Trans. Networking*, vol. 29, no. 4, pp. 1605–1619, August 2021, doi: 10.1109/TNET.2021.3073160.
[37] E. G. Larsson, O. Edfors, F. Tufvesson, and T. L. Marzetta, "Massive MIMO for next generation wireless systems," *IEEE Commun. Mag.*, vol. 52, no. 2, pp. 186–195, February 2014.
[38] Q. Yang et al., "BigStation: Enabling scalable real-time signal processing in large MU-MIMO systems," *Proc. ACM SIGCOMM Conf.*, Hong Kong, China, August 2013, pp. 1–39.
[39] C. Shepard et al., "Argos: Practical many-antenna base stations," *Proc. 18th Annu. Int. Conf. Mobile Comput. Netw.*, Istanbul, Turkey, 2012, pp. 53–64.
[40] X. Xie, E. Chai, X. Zhang, K. Sundaresan, A. Khojastepour, and S. Rangarajan, "Hekaton: Efficient and practical large-scale MIMO," *Proc. 21st Annu. Int. Conf. Mobile Comput. Netw.*, Paris, France, September 2015, pp. 304–316.
[41] Y. Ouyang, D. J. Love, and W. J. Chappell, "Body-worn distributed MIMO system," *IEEE Trans. Veh. Technol.*, vol. 58, no. 4, pp. 1752–1765, May 2009, doi: 10.1109/TVT.2008.2004491.
[42] P. Hui, C. G. Hynes, J. V. Wonterghem, and D. G. Michelson, "3D autocorrelation coefficients of dipole antenna," *Electron. Lett.*, vol. 42, no. 5, pp. 257–258, Mar. 2006.
[43] B. K. Lau, S. M. S. Ow, G. Kristensson, and A. F. Molisch, "Capacity analysis for compact MIMO systems," *Proc. IEEE Veh. Technol. Conf.*, Stockholm, Sweden, May 2005, vol. 1, pp. 165–170.
[44] C. Waldschmidt, T. Fugen, and W. Wiesbeck, "Spiral and dipole antennas for indoor MIMO-systems," *IEEE Antennas Wireless Propag. Lett.*, vol. 1, no. 9, pp. 176–178, 2002.
[45] C. A. Balanis, *Antenna Theory: Analysis and Design*, 2nd ed. Hoboken, NJ: Wiley, 1997.

Index

A

B

C

D

E

N

O

P

U

V

W

Y

Z